Quantitative Applications in the Social Sciences

A SAGE PUBLICATIONS SERIES

1. **Analysis of Variance, 2nd Edition** *Iversen/Norpoth*
2. **Operations Research Methods** *Nagel/Neef*
3. **Causal Modeling, 2nd Edition** *Asher*
4. **Tests of Significance** *Henkel*
5. **Cohort Analysis, 2nd Edition** Glenn
6. **Canonical Analysis and Factor Comparison** *Levine*
7. **Analysis of Nominal Data, 2nd Edition** *Reynolds*
8. **Analysis of Ordinal Data** *Hildebrand/Laing/Rosenthal*
9. **Time Series Analysis, 2nd Edition** *Ostrom*
10. **Ecological Inference** *Langbein/Lichtman*
11. **Multidimensional Scaling** *Kruskal/Wish*
12. **Analysis of Covariance** **Wildt/Ahtola**
13. **Introduction to Factor Analysis** *Kim/Mueller*
14. **Factor Analysis** **Kim/Mueller**
15. **Multiple Indicators** *Sullivan/Feldman*
16. **Exploratory Data Analysis** *Hartwig/Dearing*
17. **Reliability and Validity Assessment** *Carmines/Zeller*
18. **Analyzing Panel Data** *Markus*
19. **Discriminant Analysis** *Klecka*
20. **Log-Linear Models** *Knoke/Burke*
21. **Interrupted Time Series Analysis** *McDowall/McCleary/Meidinger/Hay*
22. **Applied Regression** *Lewis-Beck*
23. **Research Designs** *Spector*
24. **Unidimensional Scaling** *McIver/Carmines*
25. **Magnitude Scaling** *Lodge*
26. **Multiattribute Evaluation** *Edwards/Newman*
27. **Dynamic Modeling** *Huckfeldt/Kohfeld/Likens*
28. **Network Analysis** *Knoke/Kuklinski*
29. **Interpreting and Using Regression** *Achen*
30. **Test Item Bias** *Osterlind*
31. **Mobility Tables** *Hout*
32. **Measures of Association** *Liebetrau*
33. **Confirmatory Factor Analysis** *Long*
34. **Covariance Structure Models** *Long*
35. **Introduction to Survey Sampling** *Kalton*
36. **Achievement Testing** *Bejar*
37. **Nonrecursive Causal Models** *Berry*
38. **Matrix Algebra** *Namboodiri*
39. **Introduction to Applied Demography** *Rives/Serow*
40. **Microcomputer Methods for Social Scientists, 2nd Edition** *Schrodt*
41. **Game Theory** *Zagare*
42. **Using Published Data** *Jacob*
43. **Bayesian Statistical Inference** *Iversen*
44. **Cluster Analysis** *Aldenderfer/Blashfield*
45. **Linear Probability, Logit, and Probit Models** *Aldrich/Nelson*
46. **Event History Analysis** *Allison*
47. **Canonical Correlation Analysis** *Thompson*
48. **Models for Innovation Diffusion** *Mahajan/Peterson*
49. **Basic Content Analysis, 2nd Edition** *Weber*
50. **Multiple Regression in Practice** *Berry/Feldman*
51. **Stochastic Parameter Regression Models** *Newbold/Bos*
52. **Using Microcomputers in Research** *Madron/Tate/Brookshire*
53. **Secondary Analysis of Survey Data** *Kiecolt/Nathan*
54. **Multivariate Analysis of Variance** *Bray/Maxwell*
55. **The Logic of Causal Order** *Davis*
56. **Introduction to Linear Goal Programming** *Ignizio*
57. **Understanding Regression Analysis** *Schroeder/Sjoquist/Stephan*
58. **Randomized Response** *Fox/Tracy*
59. **Meta-Analysis** *Wolf*
60. **Linear Programming** *Feiring*
61. **Multiple Comparisons** *Klockars/Sax*
62. **Information Theory** *Krippendorff*
63. **Survey Questions** *Converse/Presser*
64. **Latent Class Analysis** *McCutcheon*
65. **Three-Way Scaling and Clustering** *Arabie/Carroll/DeSarbo*
66. **Q Methodology** *McKeown/Thomas*
67. **Analyzing Decision Making** *Louviere*
68. **Rasch Models for Measurement** *Andrich*
69. **Principal Components Analysis** *Dunteman*
70. **Pooled Time Series Analysis** *Sayrs*
71. **Analyzing Complex Survey Data, 2nd Edition** *Lee/Forthofer*
72. **Interaction Effects in Multiple Regression, 2nd Edition** *Jaccard/Turrisi*
73. **Understanding Significance Testing** *Mohr*
74. **Experimental Design and Analysis** *Brown/Melamed*
75. **Metric Scaling** *Weller/Romney*
76. **Longitudinal Research, 2nd Edition** *Menard*
77. **Expert Systems** *Benfer/Brent/Furbee*
78. **Data Theory and Dimensional Analysis** *Jacoby*
79. **Regression Diagnostics** *Fox*
80. **Computer-Assisted Interviewing** *Saris*
81. **Contextual Analysis** *Iversen*
82. **Summated Rating Scale Construction** *Spector*
83. **Central Tendency and Variability** *Weisberg*
84. **ANOVA: Repeated Measures** *Girden*
85. **Processing Data** *Bourque/Clark*
86. **Logit Modeling** *DeMaris*
87. **Analytic Mapping and Geographic Databases** *Garson/Biggs*
88. **Working With Archival Data** *Elder/Pavalko/Clipp*
89. **Multiple Comparison Procedures** *Toothaker*
90. **Nonparametric Statistics** *Gibbons*
91. **Nonparametric Measures of Association** *Gibbons*

Quantitative Applications in the Social Sciences

A SAGE PUBLICATIONS SERIES

92. **Understanding Regression Assumptions** *Berry*
93. **Regression With Dummy Variables** *Hardy*
94. **Loglinear Models With Latent Variables** *Hagenaars*
95. **Bootstrapping** *Mooney/Duval*
96. **Maximum Likelihood Estimation** *Eliason*
97. **Ordinal Log-Linear Models** *Ishii-Kuntz*
98. **Random Factors in ANOVA** *Jackson/Brashers*
99. **Univariate Tests for Time Series Models** *Cromwell/Labys/Terraza*
100. **Multivariate Tests for Time Series Models** *Cromwell/Hannan/Labys/Terraza*
101. **Interpreting Probability Models: Logit, Probit, and Other Generalized Linear Models** *Liao*
102. **Typologies and Taxonomies** *Bailey*
103. **Data Analysis: An Introduction** *Lewis-Beck*
104. **Multiple Attribute Decision Making** *Yoon/Hwang*
105. **Causal Analysis With Panel Data** *Finkel*
106. **Applied Logistic Regression Analysis, 2nd Edition** *Menard*
107. **Chaos and Catastrophe Theories** *Brown*
108. **Basic Math for Social Scientists: Concepts** *Hagle*
109. **Basic Math for Social Scientists: Problems and Solutions** *Hagle*
110. **Calculus** *Iversen*
111. **Regression Models: Censored, Sample Selected, or Truncated Data** *Breen*
112. **Tree Models of Similarity and Association** *Corter*
113. **Computational Modeling** *Taber/Timpone*
114. **LISREL Approaches to Interaction Effects in Multiple Regression** *Jaccard/Wan*
115. **Analyzing Repeated Surveys** *Firebaugh*
116. **Monte Carlo Simulation** *Mooney*
117. **Statistical Graphics for Univariate and Bivariate Data** *Jacoby*
118. **Interaction Effects in Factorial Analysis of Variance** *Jaccard*
119. **Odds Ratios in the Analysis of Contingency Tables** *Rudas*
120. **Statistical Graphics for Visualizing Multivariate Data** *Jacoby*
121. **Applied Correspondence Analysis** *Clausen*
122. **Game Theory Topics** *Fink/Gates/Humes*
123. **Social Choice: Theory and Research** *Johnson*
124. **Neural Networks** *Abdi/Valentin/Edelman*
125. **Relating Statistics and Experimental Design: An Introduction** *Levin*
126. **Latent Class Scaling Analysis** *Dayton*
127. **Sorting Data: Collection and Analysis** *Coxon*
128. **Analyzing Documentary Accounts** *Hodson*
129. **Effect Size for ANOVA Designs** *Cortina/Nouri*
130. **Nonparametric Simple Regression: Smoothing Scatterplots** *Fox*
131. **Multiple and Generalized Nonparametric Regression** *Fox*
132. **Logistic Regression: A Primer** *Pampel*
133. **Translating Questionnaires and Other Research Instruments: Problems and Solutions** *Behling/Law*
134. **Generalized Linear Models: A Unified Approach** *Gill*
135. **Interaction Effects in Logistic Regression** *Jaccard*
136. **Missing Data** *Allison*
137. **Spline Regression Models** *Marsh/Cormier*
138. **Logit and Probit: Ordered and Multinomial Models** *Borooah*
139. **Correlation: Parametric and Nonparametric Measures** *Chen/Popovich*
140. **Confidence Intervals** *Smithson*
141. **Internet Data Collection** *Best/Krueger*
142. **Probability Theory** *Rudas*
143. **Multilevel Modeling** *Luke*
144. **Polytomous Item Response Theory Models** *Ostini/Nering*
145. **An Introduction to Generalized Linear Models** *Dunteman/Ho*
146. **Logistic Regression Models for Ordinal Response Variables** *O'Connell*
147. **Fuzzy Set Theory: Applications in the Social Sciences** *Smithson/Verkuilen*
148. **Multiple Time Series Models** *Brandt/Williams*
149. **Quantile Regression** *Hao/Naiman*
150. **Differential Equations: A Modeling Approach** *Brown*
151. **Graph Algebra: Mathematical Modeling With a Systems Approach** *Brown*
152. **Modern Methods for Robust Regression** *Andersen*
153. **Agent-Based Models** *Gilbert*
154. **Social Network Analysis, 2nd Edition** *Knoke/Yang*
155. **Spatial Regression Models** *Ward/Gleditsch*
156. **Mediation Analysis** *Iacobucci*
157. **Latent Growth Curve Modeling** *Preacher/Wichman/MacCallum/Briggs*
158. **Introduction to the Comparative Method With Boolean Algebra** *Caramani*
159. **A Mathematical Primer for Social Statistics** *Fox*
160. **Fixed Effects Regression Models** *Allison*
161. **Differential Item Functioning, 2nd Edition** *Osterlind/Everson*
162. **Quantitative Narrative Analysis** *Franzosi*
163. **Multiple Correspondence Analysis** *LeRoux/Rouanet*
164. **Association Models** *Wong*
165. **Fractal Analysis** *Brown/Liebovitch*
166. **Assessing Inequality** *Hao/Naiman*
167. **Graphical Models and the Multigraph Representation for Categorical Data** *Khamis*
168. **Nonrecursive Models** *Paxton/Hipp/Marquart-Pyatt*

Series/Number 08-168

NONRECURSIVE MODELS

Endogeneity, Reciprocal Relationships, and Feedback Loops

Pamela Paxton
The University of Texas at Austin

John R. Hipp
University of California, Irvine

Sandra Marquart-Pyatt
Michigan State University

Los Angeles | London | New Delhi
Singapore | Washington DC

Los Angeles | London | New Delhi
Singapore | Washington DC

FOR INFORMATION:

SAGE Publications, Inc.
2455 Teller Road
Thousand Oaks, California 91320
E-mail: order@sagepub.com

SAGE Publications Ltd.
1 Oliver's Yard
55 City Road
London EC1Y 1SP
United Kingdom

SAGE Publications India Pvt. Ltd.
B 1/I 1 Mohan Cooperative Industrial Area
Mathura Road, New Delhi 110 044
India

SAGE Publications Asia-Pacific Pte. Ltd.
33 Pekin Street #02-01
Far East Square
Singapore 048763

Executive Editor: Vicki Knight
Associate Editor: Lauren Habib
Editorial Assistant: Kalie Koscielak
Production Editor: Brittany Bauhaus
Permissions Editor: Adele Hutchinson
Copy Editor: QuADS Prepress (P) Ltd.
Typesetter: C&M Digitals (P) Ltd.
Proofreader: Scott Oney
Indexer: Diggs Publication Services, Inc.
Cover Designer: Candice Harman
Marketing Manager: Helen Salmon

Printed in the United States of America

Library of Congress Cataloging-in-Publication Data

Paxton, Pamela.

Nonrecursive models : endogeneity, reciprocal relationships, and feedback loops / Pamela Paxton, John R. Hipp, Sandra Marquart-Pyatt.

p. cm. — (Quantitative applications in the social sciences ; 168)
Includes bibliographical references and index.

ISBN 978-1-4129-7444-8 (pbk.)

1. Social sciences—Mathematical models. I. Hipp, John R. II. Marquart-Pyatt, Sandra T. III. Title.

H61.25.P398 2011

300.1'5129426—dc22 2010048517

This book is printed on acid-free paper.

11 12 13 14 15 10 9 8 7 6 5 4 3 2 1

CONTENTS

ABOUT THE AUTHORS

Pamela Paxton is professor of Sociology and Government and Christine and Stanley E. Adams, Jr. Centennial Professor in the Liberal Arts at the University of Texas at Austin. She has taught at the Inter-University Consortium for Political and Social Research (ICPSR) Summer Training Program in Advanced Statistical Techniques and has consulted for the U.S. Agency for International Development (USAID). She is the author of articles and books on prosocial behavior, women in politics, and quantitative methodology. She received her undergraduate degree in economics and sociology from the University of Michigan and her PhD in sociology from the University of North Carolina at Chapel Hill.

John R. Hipp is an associate professor in the Departments of Criminology, Law and Society, and Sociology at the University of California Irvine. He currently teaches a graduate course on structural equation models, and he taught a course on simultaneous equation models for the Inter-University Consortium for Political and Social Research (ICPSR) Summer Program in Quantitative Methods for 3 years, from 2003 to 2005. His substantive research interests focus on how neighborhoods change over time, how that change both affects and is affected by neighborhood crime, and the role networks and institutions play in that change. He approaches these questions using quantitative methods. He has published methodological work in journals such as *Sociological Methodology*, *Psychological Methods*, and *Structural Equation Modeling* and a book chapter in the *New Handbook on Data Analysis* (edited by M. A. Hardy), and he contributed to an entry in the *Encyclopedia of Social Science Research Methods* (edited by M. Lewis-Beck, A. Bryan, and T. F. Liao). He has published substantive work in journals such as *American Sociological Review*, *Criminology*, *Social Forces*, *Social Problems*, *Mobilization*, *City & Community*, *Urban Studies*, and *Journal of Urban Affairs*.

Sandra Marquart-Pyatt is an assistant professor in Sociology and the Environmental Science and Policy Program at Michigan State University. She has been an instructor for the Inter-University Consortium for Political and Social Research (ICPSR) Summer Program in Quantitative Methods at the University of Michigan for a simultaneous equation models course and regularly teaches graduate courses on structural equation modeling. Her research and teaching areas of expertise are in comparative social change, environmental sociology, political sociology, and quantitative methods. She is the author of work on cross-national patterns on an array of environmental attitudes and behaviors, comparative work on environmental sustainability, and democratic values cross-nationally. She received her PhD in sociology from The Ohio State University.

SERIES EDITOR'S INTRODUCTION

Linear structural equation models (SEMs) are multi-equation linear regression models in which the response (or dependent) variable in one regression equation can appear as an explanatory (independent) variable in another. The models are termed *structural* because their aspiration is typically to estimate structural, or causal, relationships. Because current applications of SEMs usually entail latent variables, Paxton, Hipp, and Marquart-Pyatt employ the strictly synonymous term *simultaneous equation models* to describe SEMs in which there are no latent variables except for structural disturbances (i.e., errors in structural equations rather than measurement errors in observed variables). These observed-variable simultaneous equation models are the subject of their monograph. They focus, in particular, on nonrecursive linear simultaneous equation models, in which some of the explanatory variables in a structural equation may be correlated with the disturbance of that equation. Nonrecursive models cannot be estimated consistently by ordinary least squares regression.

From their origins in several disciplines, simultaneous equation models became generally popular in the social sciences in the late 1960s and 1970s, and SEMs are currently in wide use in a number of disciplines. I have been repeatedly struck, however, by the shallowness in the understanding of these models by the large majority of their users, and even, on occasion, by people who have published on the topic. For example, many practitioners of structural equation modeling appear not to understand the identification problem—how to determine whether or not a model that has been specified can be estimated from data.

The current monograph on nonrecursive simultaneous equation models serves to refocus attention on fundamental issues in structural equation modeling, including the specification, identification, estimation, assessment, and interpretation of the models. The authors deal with important topics that are typically neglected, such as limited-information estimation methods (as opposed to full-information methods such as multivariate-normal full-information maximum likelihood), most notably two-stage least squares, and the quality of instrumental variables.

It is my expectation that most users of simultaneous equation models—both in the narrow sense of that term employed in this monograph and in the broader sense of SEMs that include latent variables—will learn a great deal of value from this monograph.

Editor's note: This monograph was begun under the direction of the previous series editor, Tim Futing Liao.

—*John Fox*
Series Editor

ACKNOWLEDGMENTS

The authors would like to thank the ICPSR summer program, Hank Heitowit, and our ICPSR students throughout the years for inspiring this book. We also thank Tim Futing Liao and John Fox.

The authors and SAGE would like to acknowledge the contributions of the following reviewers:

Kenneth A. Bollen, *University of North Carolina at Chapel Hill*

Dawn Iacobucci, *Owen Graduate School of Management, Vanderbilt University*

William G. Jacoby, *Michigan State University* and *ICPSR*

David McDowall, *School of Criminal Justice, University at Albany–SUNY*

CHAPTER 1. INTRODUCTION

Throughout the social sciences, scholars are aware that models estimated using observational data are likely affected by endogeneity for one or more predictors. Indeed, social scientists often construct models that explicitly specify reciprocal relationships or feedback loops among multiple outcomes. But many social scientists may underestimate the consequences of endogeneity resulting from such "nonrecursive models" or be unaware of how to correctly model or otherwise address it. As a result, it is not unusual for researchers to adopt an estimation strategy that effectively ignores such possible endogeneity and results in biased estimates. Equally problematic is the estimation or interpretation of results from nonrecursive models in which the assumptions of the model are either statistically or theoretically inappropriate.

This monograph provides an overview of methods appropriate for the analysis of nonrecursive simultaneous equation models. Simultaneous equation models have at least two equations and stand in contrast to the more common instance in which social scientists posit a single equation with a single outcome. A simultaneous equation model is nonrecursive if (1) two of the outcomes in the model affect one another (a reciprocal relationship) or there is a feedback loop at some point in the system of equations (i.e., a causal path can be traced from one variable back to itself), and/or (2) at least some disturbances are correlated. A model is termed fully recursive if there is no posited reciprocal relationship or feedback loop, *and* there are assumed to be no relationships among the error terms of the equations.

We introduce the specification, identification, and estimation of simultaneous equation models, how to assess the quality of the estimates, and how to correctly interpret results, with a focus on nonrecursive models. In nonrecursive models, identification of an equation (demonstrating that unique values for the parameters can be estimated) often requires the incorporation of a variable that is correlated with the problematic variable but uncorrelated with the disturbance term of the equation: This is referred to as an instrumental variable. Another focus of the monograph will be introducing proper selection, use, and assessment of instrumental variables. We will emphasize that properly selecting instrumental variables is important regardless of which estimator is used.

Throughout the monograph, we blend two complementary perspectives on simultaneous equation nonrecursive models. First, simultaneous equations are addressed in the structural equation modeling (SEM) with latent variables literature (e.g., Bollen, 1989b; Kaplan, 2009).[1] The SEM literature stresses specification using path diagrams, full-information estimation, and global assessment of model fit. The SEM approach to simultaneous equations is limited, however, by a relative neglect of nonrecursive models, a lack of assessment of individual equations, and little discussion of the quality of the instrumental variables used to identify nonrecursive equations. Many SEM treatments of simultaneous equations focus almost exclusively on full-information estimation strategies such as maximum likelihood. But a limited-information approach is useful for several tests that assess the quality of the individual equations in the model. A focus on maximum likelihood alone leaves researchers without the tools to properly evaluate the assumptions underlying their nonrecursive models.

A complementary approach to simultaneous equation models comes from the econometrics tradition (e.g., Greene, 2008; Kennedy, 2008; Wooldridge, 2002, 2009).[2] The econometric literature highlights the link between simultaneous equation models and assumption violations of traditional regression techniques, stresses the identification of nonrecursive models using instrumental variables, and focuses on limited-information estimators to a greater extent. But because the econometrics literature focuses on individual equations, it does not stress the interpretation of results as part of a multiequation model. Furthermore, the range of possible assessments of instrumental variables are rarely consolidated and compared.

Throughout this monograph, we take the position that limited-information estimators such as two-stage least squares are not outdated methods that can be safely ignored by researchers using structural equation software packages. Instead, we argue that a need for a clear and rigorous understanding of nonrecursive models has developed. With increasing ease of software programs, social scientists can now estimate nonrecursive models without fully understanding their unique attributes. Many full-information software programs do not provide equation-by-equation assessment of the quality of the model proposed. Important information is contained in the reduced-form equations of such models, and this information is often opaque to

---

[1]Simultaneous equation models differ from SEMs in that measurement error in the variables is rarely considered in simultaneous equation models, and there is no attempt to estimate latent variables (see Bollen, 1989b, pp. 80–150). The distinction is not hard and fast and can be viewed as largely a matter of emphasis.

[2]Confusion can arise from terminology. For example, some economists will refer to simultaneous equation models as we define them as structural equation models, although their approach is different from the latent variable SEM approach.

researchers adopting a full-information estimation approach. By integrating the econometrics and SEM approaches to simultaneous equation models, this monograph provides a "back to basics" approach that is directly relevant to researchers wishing to estimate nonrecursive models. The monograph is oriented around five steps we advocate for modeling: specification, identification, estimation, assessment, and interpretation. Related volumes in the series expand on particular steps: for example, Berry (1984) provides extensive coverage of identification.

In our presentation, we assume knowledge of multiple regression analysis. Readers will further benefit if they have some familiarity with SEM. Good presentations of SEM are available in Bollen (1989b) and Kaplan (2009). Models will be presented and discussed as path diagrams, equations, or matrix equations. Good introductions to matrix algebra can be found in Gill (2006, chaps. 3 and 4), Fox (2009), and Namboodiri (1984). We present examples using some of the available software, with a focus on software such as SAS and Stata that allow estimation of individual equations.

CHAPTER 2. SPECIFICATION OF SIMULTANEOUS EQUATION MODELS

In model specification, the researcher uses prior theory to detail a series of equations and represent these using path models, equations, and/or matrices. Simultaneous equation models contain random variables (i.e., observed variables and error terms) and structural parameters (i.e., constants providing intercepts and the relationships between variables). The variables of a simultaneous equation model may be linked through direct relationships, indirect relationships, reciprocal relationships, feedback loops, and/or correlations between disturbances. Theory plays an instrumental role in model creation and determines the theoretical, or structural, relationships between the variables of interest. Empirical methods assess the fit of these specified models to the data, as we will explain later.

The general matrix representation of simultaneous equation models appears in Equation 2.1:

$$\mathbf{y} = \mathbf{B}\mathbf{y} + \mathbf{\Gamma}\mathbf{x} + \boldsymbol{\zeta} \tag{2.1}$$

Endogenous variables, denoted by **y**, are outcome variables or variables determined within the model. The vector of endogenous variables has dimensions $p \times 1$. Exogenous variables, denoted with the $q \times 1$ vector, **x**, are exogenous variables in the model (i.e., they are not explained by the model). For computational ease, random variables are assumed to be deviated from their means. Disturbances, or errors in the equations, are represented with $\boldsymbol{\zeta}$, a $p \times 1$ vector. There is one disturbance per endogenous variable, hence the similar dimensions. Gamma coefficients (γ) describe the effect of exogenous variables on the endogenous variables and are summarized in the coefficient matrix $\mathbf{\Gamma}$, which has dimensions $p \times q$. Beta coefficients (β) describe the effect of an endogenous variable on another endogenous variable. They are summarized in the coefficient matrix **B**, with dimensions $p \times p$. Table 2.1 summarizes the components of a simultaneous equation model, including vector/matrix names, definitions, and dimensions (Bollen, 1989b; Kaplan, 2009).

Also in Table 2.1, two additional covariance matrices are important in describing simultaneous equation models. The matrix designated by $\mathbf{\Phi}$ is the variance/covariance matrix of the exogenous variables (xs), and $\mathbf{\Psi}$ is the covariance matrix of the disturbance terms (ζs). Both covariance matrices are symmetric.

Table 2.1 Notation for Simultaneous Equation Models

Vector/Matrix	*Definition*	*Dimensions*
Variables		
$\mathbf{y}$	Endogenous variables	$p \times 1$
$\mathbf{x}$	Exogenous variables	$q \times 1$
$\boldsymbol{\zeta}$	Disturbance terms or errors in equations	$p \times 1$
Coefficients		
$\boldsymbol{\Gamma}$	Coefficient matrix for the exogenous variables; the effect of exogenous variables on endogenous variables; direct effects of x on y	$p \times q$
$\mathbf{B}$	Coefficient matrix for the endogenous variables; the effect of endogenous variables on endogenous variables; direct effects of y on y	$p \times p$
Covariance matrices		
$\boldsymbol{\Phi}$	Covariance matrix of exogenous variables, x	$q \times q$
$\boldsymbol{\Psi}$	Covariance matrix of disturbance terms, ζ	$p \times p$

A number of assumptions apply to these models. Simultaneous equation models assume that the endogenous and exogenous variables are directly measured and have no measurement error. The disturbances include all variables influencing y that are omitted from the equation and are assumed to have expected values of zero ($E(\zeta) = 0$). Disturbances are further assumed to be uncorrelated with the exogenous variables, homoscedastic and nonautocorrelated. Violations are possible; some will be treated in this monograph, while others are addressed elsewhere (e.g., Kmenta, 1997). The random variables are assumed not to have instantaneous effects on themselves.

Path Diagrams, Equations, Matrices: An Example of Specification

A path diagram pictorially represents how variables are related to one another in a theoretical model. Path diagrams use particular conventions: Variables shown in rectangles are observed variables, single-headed arrows denote the direction of influence, and double-headed arrows depict a covariance not explained in the model. Errors in the equations could technically be placed in ovals (representing unobserved or latent variables), but they are typically depicted unenclosed. Figure 2.1 is a path diagram of a recursive model containing two exogenous and two endogenous variables. Following the notation introduced above, the gamma coefficients represent direct effects of exogenous (x) variables on endogenous (y) variables: γ_{11} is the coefficient for the path from x_1 to y_1, γ_{12} is the coefficient for the path from x_2 to y_1, and γ_{22} is the coefficient for the path from x_2 to y_2.[1] β_{21} is the coefficient for the path from y_1 to y_2.[2]

Figure 2.1 Path Diagram of a Recursive Model

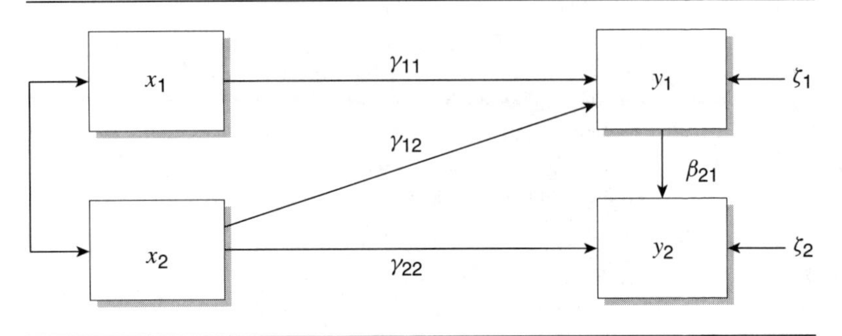

The model can be written as a series of equations, one for each endogenous variable. Two equations correspond to the path diagram in Figure 2.1 (intercepts are unnecessary, given that the random variables are assumed to be deviated from their means).

$$y_1 = \gamma_{11}x_1 + \gamma_{12}x_2 + \zeta_1 \tag{2.2}$$

$$y_2 = \beta_{21}y_1 + \gamma_{22}x_2 + \zeta_2 \tag{2.3}$$

The two representations are equivalent, although the equations do not provide the important information that ζ_1 and ζ_2 are uncorrelated.

[1]Coefficient subscripts follow particular conventions: The first number in the subscript denotes the variable being influenced, and the second number is the variable doing the influencing. Subscripts also denote placement in the coefficient matrix, row and column positions.

[2]The path with two-headed arrows between the exogenous variables x_1 and x_2 represents the observed covariance between these two variables. Covariances are generally represented as two-headed arrows: One convention is to use a curved path, and another convention is to use a straight line with multiple-headed arrows.

The same model can finally be written as a matrix equation:

$$\begin{bmatrix} y_1 \\ y_2 \end{bmatrix} = \begin{bmatrix} 0 & 0 \\ \beta_{21} & 0 \end{bmatrix} \begin{bmatrix} y_1 \\ y_2 \end{bmatrix} + \begin{bmatrix} \gamma_{11} & \gamma_{12} \\ 0 & \gamma_{22} \end{bmatrix} \begin{bmatrix} x_1 \\ x_2 \end{bmatrix} + \begin{bmatrix} \zeta_1 \\ \zeta_2 \end{bmatrix} \tag{2.4}$$

With two endogenous variables in the model, the **y** vector has dimensions 2 × 1. The **B** matrix, the coefficient matrix for endogenous variables, includes β_{21}, the effect of y_1 on y_2. The **Γ** matrix contains coefficients for the effects of exogenous variables on endogenous variables and has dimensions 2 × 2. The two disturbance terms are shown in **ζ**, written as a vector with dimensions 2 × 1.

The variances of the exogenous variables and the assumed covariance between them are shown in the **Φ** matrix with dimensions 2 × 2.[3] We follow a common convention of only displaying the lower diagonal of symmetric matrices. The **Ψ** matrix, which includes variances of the errors in the equations, has dimensions 2 × 2. In this model, the **Ψ** matrix is diagonal—the disturbances are not correlated.

$$\mathbf{\Phi} = \begin{bmatrix} \phi_{11} & \\ \phi_{21} & \phi_{22} \end{bmatrix}$$

$$\mathbf{\Psi} = \begin{bmatrix} \psi_{11} & \\ 0 & \psi_{22} \end{bmatrix} \tag{2.5}$$

The matrix equations reveal certain properties of simultaneous equation models. For instance, variables do not have instantaneous effects on themselves, as shown in the zeros down the diagonal of the **B** matrix.

From Theory to Models: The Implied Covariance Matrix

Understanding simultaneous equation models is aided by switching one's frame of reference to a focus on the covariances among variables rather than individual cases in a sample (Bollen, 1989b).[4] The focus is on the fundamental statistical hypothesis,

$$\mathbf{\Sigma} = \mathbf{\Sigma}(\boldsymbol{\theta}) \tag{2.6}$$

---

[3]Unless a researcher is working with experimental data, in which zero correlation between exogenous variables can be assured, it is typical not to assume zero correlation.

[4]Certainly individual observations matter a great deal and cannot be ignored. For example, outliers may influence the results of a given analysis.

where $\mathbf{\Sigma}$ is the population covariance matrix of the observed variables, $\boldsymbol{\theta}$ is a vector of the parameters to be estimated, and $\mathbf{\Sigma}(\boldsymbol{\theta})$ is the covariance matrix implied by your model (written as a function of the model parameters). In layman's terms, Equation 2.6 equates "your data" (at the population level) and "your model," which is precisely what conventional statistical techniques are designed to do. The fundamental statistical hypothesis underlies all aspects of modeling in simultaneous equation models: specification, identification, estimation, and assessment.

To understand how $\mathbf{\Sigma} = \mathbf{\Sigma}(\boldsymbol{\theta})$ relates to specification, it is useful to take a step back and view the big picture. Researchers typically have a set of variables in which they are interested, and they have some model in mind of how these variables fit together. This model may appear in a set of equations or as a path diagram.

The bulk of information needed for specification and estimation is summarized in the variances and covariances between the observed variables. That is, the total raw association between the variables is captured in the matrix $\mathbf{\Sigma}$ and is known at least in the population.

Figure 2.2 Three Models From the Same Variables

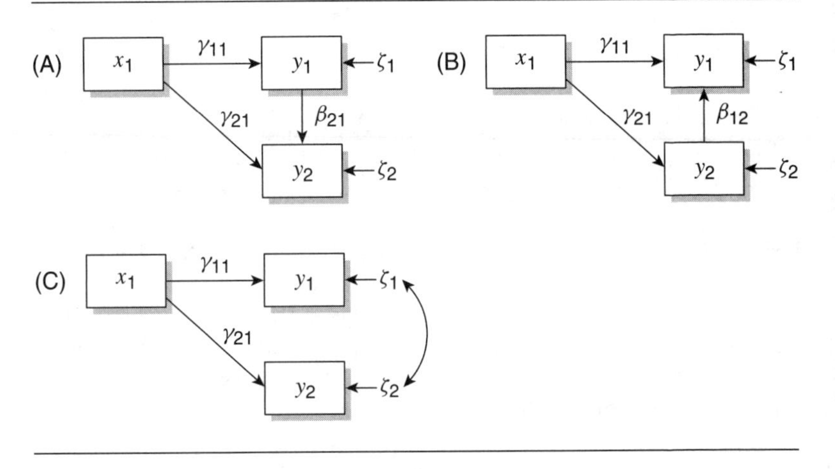

A researcher's model makes the argument that the association between variables is due to the hypothesized structure. Consider the three models displayed in Figure 2.2, Panels A, B, and C. Each model in Figure 2.2 uses the same three variables: two endogenous variables (y_1 and y_2) and one exogenous variable (x_1). But each is configured differently.

In Figure 2.2, Panel A, the endogenous variables, y_1 and y_2, are mutually dependent on the exogenous variable, x_1, and y_1 influences y_2. In Figure 2.2, Panel B, the two endogenous variables, y_1 and y_2, are again mutually dependent on the exogenous variable, x_1, but here y_2 influences y_1. In Figure 2.2, Panel C, the endogenous variables, y_1 and y_2, are mutually dependent on the single exogenous variable x_1. They are also related through an unexplained association between their errors. Theory drives which model a researcher would choose.

Creating the implied covariance matrix, $\mathbf{\Sigma(\theta)}$, allows a researcher to break down exactly how a hypothesized model relates to the known variances and covariances among the observed variables. As an example, consider the covariance between x_1 and y_1: $\text{COV}(x_1, y_1)$. This is a known quantity in the population; there is a known association between these two variables. What does the model in Figure 2.2, Panel A, imply about this association? The answer can be discovered by substituting the equations of the model for x_1 and y_1,

$$\text{COV}(x_1, y_1) = \text{COV}(x_1, \gamma_{11}x_1 + \zeta_1) \tag{2.7}$$

Covariance algebra aids in rearranging terms so that Equation 2.7 becomes[5]

$$\text{COV}(x_1, y_1) = \text{COV}(x_1, \gamma_{11}x_1) + \text{COV}(x_1, \zeta_1)$$

Exogenous variables are assumed to be uncorrelated with the disturbances, leaving

$$\text{COV}(x_1, y_1) = \gamma_{11}\text{VAR}(x_1)$$

Finally, the variance of the exogenous variable, x_1, is a parameter to be estimated in the model. It appears in the $\mathbf{\Phi}$ matrix. Making this explicit,

$$\text{COV}(x_1, y_1) = \gamma_{11}\phi_{11} \tag{2.8}$$

Equation 2.8 provides the model-implied covariance for $\text{COV}(x_1, y_1)$. In the model in Figure 2.2, Panel A, the covariance between x_1 and y_1 is

[5]The following rules and definitions are used (Bollen, 1989b, p. 21). Defining c as a constant, and x_1, x_2, and x_3 as random variables: (1) $\text{COV}(c, x_1) = 0$, (2) $\text{COV}(cx_1, x_2) = c\text{COV}(x_1, x_2)$, and (3) $\text{COV}(x_1 + x_2, x_3) = \text{COV}(x_1, x_3) + \text{COV}(x_2, x_3)$. Note also that $\text{VAR}(x_1) = \text{COV}(x_1, x_1)$.

implied by the model to be due to the variance of $x_1(\phi_{11})$ and its effect on $y_1(\gamma_{11})$.

Of course, COV(x_1, y_1) is only one of the six possible variances and covariances between the three observed variables. The full covariance matrix of the observed variables is

$$\mathbf{\Sigma} = \begin{bmatrix} \text{VAR}(y_1) & & \\ \text{COV}(y_2, y_1) & \text{VAR}(y_2) & \\ \text{COV}(x_1, y_1) & \text{COV}(x_1, y_2) & \text{VAR}(x_1) \end{bmatrix}$$

For this particular model, there are six variance and covariance elements in the population, and each can be written as a function of the theoretical model appearing in Figure 2.2, Panel A. To provide one more example, consider the covariance of x_1 with y_2. Again, we substitute the equations of the theoretical model to determine what it implies about this covariance.

$$\text{COV}(x_1, y_2) = \text{COV}(x_1, \beta_{21}y_1 + \gamma_{21}x_1 + \zeta_2) \tag{2.9}$$

The variable y_1 is endogenous and requires further substitution:

$$\begin{aligned} \text{COV}(x_1, y_2) &= \text{COV}(x_1, \beta_{21}(\gamma_{11}x_1 + \zeta_1) + \gamma_{21}x_1 + \zeta_2) \\ &= \text{COV}(x_1, \beta_{21}\gamma_{11}x_1) + \text{COV}(x_1, \beta_{21}\zeta_1) \\ &\quad + \text{COV}(x_1, \gamma_{21}x_1) + \text{COV}(x_1, \zeta_2) \\ &= \beta_{21}\gamma_{11}\phi_{11} + \gamma_{21}\phi_{11} \end{aligned} \tag{2.10}$$

In short, the theoretical model implies that the relationship between x_1 and y_2 is due to both the direct effect of x_1 on y_2 and the indirect effect of x_1 on y_2 through y_1. Decomposing covariances in this way shows how the model specified in the path diagram and equations relates to the observed covariances in $\mathbf{\Sigma}$. Four implied covariances remain to be calculated to complete the model-implied covariance matrix.

$$\begin{aligned} \text{COV}(y_1, y_2) &= \gamma_{11}^2\beta_{21}\phi_{11} + \gamma_{11}\gamma_{21}\phi_{11} + \beta_{21}\psi_{11} \\ \text{VAR}(x_1) &= \text{COV}(x_1, x_1) = \phi_{11} \\ \text{VAR}(y_1) &= \gamma_{11}^2\phi_{11} + \psi_{11} \\ \text{VAR}(y_2) &= (\beta_{21}^2\gamma_{11}^2\phi_{11} + 2\beta_{21}\gamma_{11}\gamma_{21} + \gamma_{21}^2)\phi_{11} + \beta_{21}^2\psi_{11} + \psi_{22} \end{aligned}$$

Pulling everything together, for the three-variable model in Figure 2.2, Panel A, the observed covariance matrix, $\mathbf{\Sigma}$, is

$$\mathbf{\Sigma} = \begin{bmatrix} \mathrm{VAR}(y_1) & & \\ \mathrm{COV}(y_2, y_1) & \mathrm{VAR}(y_2) & \\ \mathrm{COV}(x_1, y_1) & \mathrm{COV}(x_1, y_2) & \mathrm{VAR}(x_1) \end{bmatrix} \qquad (2.11)$$

and the covariance matrix that is implied by the model, $\mathbf{\Sigma(\theta)}$, is

$$\mathbf{\Sigma(\theta)} = \begin{bmatrix} \gamma_{11}{}^2\phi_{11} + \psi_{11} & & \\ (\gamma_{11}{}^2\beta_{21} + \gamma_{11}\gamma_{21})\phi_{11} + \beta_{21}\psi_{11} & (\beta_{21}{}^2\gamma_{11}{}^2\phi_{11} + 2\beta_{21}\gamma_{11}\gamma_{21} + \gamma_{21}{}^2)\phi_{11} + \beta_{21}{}^2\psi_{11} + \psi_{22} & \\ \gamma_{11}\phi_{11} & \beta_{21}\gamma_{11}\phi_{11} + \gamma_{21}\phi_{11} & \phi_{11} \end{bmatrix} \qquad (2.12)$$

The fundamental statistical hypothesis, $\mathbf{\Sigma} = \mathbf{\Sigma(\theta)}$, means that each element of Equation 2.11 above is equivalent to its counterpart in Equation 2.12. This relationship between $\mathbf{\Sigma}$ and $\mathbf{\Sigma(\theta)}$ is used throughout the rest of the monograph.

To reiterate, the elements of $\mathbf{\Sigma(\theta)}$, the model-implied covariance matrix, are a function of the researcher's theoretical model as it appears in the path diagram, equations, and matrices. If we were to calculate the implied covariance matrix for the model in Figure 2.2, Panel B, it would be different from Equation 2.12. The hypothesized model changes, so the implied covariance matrix changes as well.

As will be demonstrated throughout the rest of the monograph, the relationship between $\mathbf{\Sigma}$ and $\mathbf{\Sigma(\theta)}$ is critical to all steps in modeling a simultaneous equation model. For example, researchers can sometimes use $\mathbf{\Sigma} = \mathbf{\Sigma(\theta)}$ to solve for the unknown model parameters during identification.

The Implied Covariance Matrix of a Simple Regression

For illustration, we show how to determine the implied covariance matrix of a simple regression model of one exogenous and one endogenous variable:

$$y_1 = \gamma_{11}x_1 + \zeta_1 \qquad (2.13)$$

For this model, there are three variances and covariances at the population level, shown in $\mathbf{\Sigma}$:

$$\mathbf{\Sigma} = \begin{bmatrix} \mathrm{VAR}(y_1) & \\ \mathrm{COV}(x_1, y_1) & \mathrm{VAR}(x_1) \end{bmatrix} \tag{2.14}$$

To determine the model-implied covariance matrix, $\mathbf{\Sigma(\theta)}$, for a simple regression model, we begin by solving for VAR(x_1), which can be rewritten as COV(x_1, x_1). The variance of the exogenous variable in this model, x_1, as noted earlier, is a parameter to be estimated in the model and appears in the $\mathbf{\Phi}$ matrix.

$$\mathrm{COV}(x_1, x_1) = \phi_{11} \tag{2.15}$$

Moving next to COV(x_1, y_1), we substitute for y_1, using Equation 2.13.

$$\begin{aligned} \mathrm{COV}(x_1, y_1) &= \mathrm{COV}(x_1, \gamma_{11}x_1 + \zeta_1) \\ &= \gamma_{11}\mathrm{VAR}(x_1) \\ &= \gamma_{11}\phi_{11} \end{aligned} \tag{2.16}$$

Next, we determine the variance of y_1, or COV(y_1, y_1). First, substitute for y_1:

$$\mathrm{COV}(y_1, y_1) = \mathrm{COV}(\gamma_{11}x_1 + \zeta_1, \gamma_{11}x_1 + \zeta_1)$$

Using common rules of covariance algebra yields

$$\begin{aligned} \mathrm{COV}(y_1, y_1) = {} & \mathrm{COV}(\gamma_{11}x_1, \gamma_{11}x_1) + \mathrm{COV}(\gamma_{11}x_1, \zeta_1) + \\ & \mathrm{COV}(\zeta_1, \gamma_{11}x_1) + \mathrm{COV}(\zeta_1, \zeta_1) \end{aligned}$$

Assuming exogenous variables are uncorrelated with the disturbance leaves

$$\begin{aligned} \mathrm{COV}(y_1, y_1) &= \mathrm{COV}(\gamma_{11}x_1, \gamma_{11}x_1) + \mathrm{COV}(\zeta_1, \zeta_1) \\ &= \gamma_{11}^2\phi_{11} + \psi_{11} \end{aligned} \tag{2.17}$$

Putting the elements together yields the following:

$$\mathbf{\Sigma} = \mathbf{\Sigma(\theta)}$$

$$\begin{bmatrix} \mathrm{VAR}(y_1) & \\ \mathrm{COV}(x_1, y_1) & \mathrm{VAR}(x_1) \end{bmatrix} = \begin{bmatrix} \gamma_{11}{}^2\phi_{11} + \psi_{11} & \\ \gamma_{11}\phi_{11} & \phi_{11} \end{bmatrix} \tag{2.18}$$

The fundamental statistical hypothesis, $\boldsymbol{\Sigma} = \boldsymbol{\Sigma}(\boldsymbol{\theta})$, means that each element of $\boldsymbol{\Sigma}$ is equivalent to its counterpart in $\boldsymbol{\Sigma}(\boldsymbol{\theta})$.

$$\begin{aligned} \mathrm{VAR}(y_1) &= \gamma_{11}^2\phi_{11} + \psi_{11} \\ \mathrm{COV}(x_1, y_1) &= \gamma_{11}\phi_{11} \\ \mathrm{VAR}(x_1) &= \phi_{11} \end{aligned}$$

We can solve for the regression coefficient, γ_{11},

$$\begin{aligned} \gamma_{11} &= \frac{\mathrm{COV}(x_1, y_1)}{\phi_{11}} \\ &= \frac{\mathrm{COV}(x_1, y_1)}{\mathrm{VAR}(x_1)} \end{aligned}$$

producing the well-known equation available in any introductory econometrics textbook.

Recursive and Nonrecursive Models

Simultaneous equation models can be divided into two major types: recursive and nonrecursive. A recursive simultaneous equation model has no reciprocal relationships or feedback loops and no covariances among the error terms of the equations (the disturbance of one equation is uncorrelated with the disturbances of all other equations). Formally, in recursive models $\mathbf{B}$ can be written as lower triangular and $\boldsymbol{\Psi}$ is diagonal.

A simultaneous equation model is nonrecursive if (1) any of the outcomes in the model directly affect one another (a reciprocal relationship) or there is a feedback loop at some point in the system of equations (a causal path can be traced from one variable back to itself), and/or (2) at least some disturbances are correlated.

In a previous section, we introduced a simple model of two exogenous and two endogenous variables. The model was recursive as demonstrated in both the path diagram and the matrices. Examining the path diagram (Figure 2.1) reveals no reciprocal links or feedback loops. Furthermore, the errors in the equations are not correlated. Examining the matrix equations (2.4 and 2.5) also shows a recursive model: the $\mathbf{B}$ matrix can be written as lower triangular, and the $\boldsymbol{\Psi}$ matrix is diagonal.

In contrast, Figure 2.3, Panels A and B show two types of nonrecursive models. Figure 2.3, Panel A, is nonrecursive due to the reciprocal paths between y_1 and y_2 and the correlated error between ζ_1 and ζ_2 (the presence of either would be sufficient to define the model as nonrecursive). Figure 2.3,

Figure 2.3 Two Nonrecursive Models

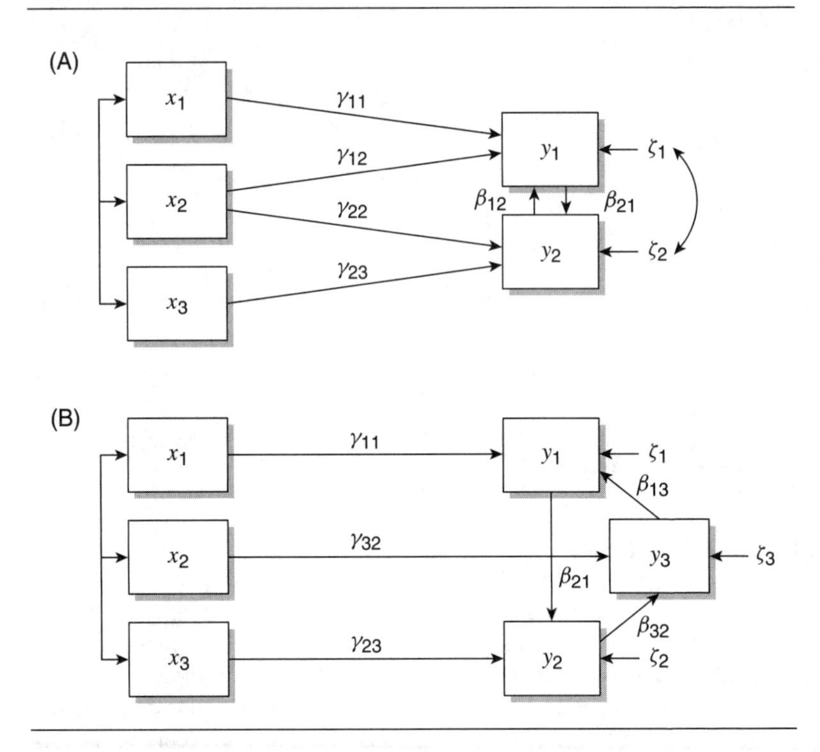

Panel B, is nonrecursive due to the presence of a feedback loop among y_1, y_2, and y_3. Note that y_1 can be traced back to itself through the change in y_2 and y_3. Similarly, y_2 and y_3 can be traced back to themselves.[6]

Focusing on Figure 2.3, Panel A, the equations of the model are as follows:

$$y_1 = \beta_{12}y_2 + \gamma_{11}x_1 + \gamma_{12}x_2 + \zeta_1 \tag{2.19}$$

$$y_2 = \beta_{21}y_1 + \gamma_{22}x_2 + \gamma_{23}x_3 + \zeta_2 \tag{2.20}$$

[6]A classic theoretical example of a three-variable feedback loop is found in climate science: A warmer climate leads to less snow and ice on the surface of the earth, which leads to less reflection of heat, which makes the climate warmer. Or consider a sleep/stress cycle when sleeping badly leads to more fatigue during the day, which makes an individual less able to cope with stressors and in turn leads to poor sleep.

The matrix equation for the model is as follows:

$$\begin{bmatrix} y_1 \\ y_2 \end{bmatrix} = \begin{bmatrix} 0 & \beta_{12} \\ \beta_{21} & 0 \end{bmatrix} \begin{bmatrix} y_1 \\ y_2 \end{bmatrix} + \begin{bmatrix} \gamma_{11} & \gamma_{12} & 0 \\ 0 & \gamma_{22} & \gamma_{23} \end{bmatrix} \begin{bmatrix} x_1 \\ x_2 \\ x_3 \end{bmatrix} + \begin{bmatrix} \zeta_1 \\ \zeta_2 \end{bmatrix} \quad (2.21)$$

and

$$\mathbf{\Phi} = \begin{bmatrix} \phi_{11} & \\ \phi_{21} & \phi_{22} \end{bmatrix}$$

$$\mathbf{\Psi} = \begin{bmatrix} \psi_{11} & \\ \psi_{12} & \psi_{22} \end{bmatrix}$$

The matrix representation helps clarify why the model is nonrecursive. First, the **B** matrix is not lower triangular and cannot be rearranged to be written as lower triangular. Furthermore, $\mathbf{\Psi}$ is not diagonal; there are off-diagonal elements. Nonrecursive models are more complicated to identify, estimate, and assess than recursive models, as will be covered in later chapters.

Direct, Indirect, and Total Effects

Simultaneous equation models contain direct, indirect, and total effects. Direct effects are effects from one variable to another variable that are not mediated by any other variable in the model. Indirect effects are paths from one variable to another that travel through at least one other variable. Total effects are the sum of direct and indirect effects, representing how much change should occur in the outcome variable for a given shift in the antecedent variable.

Specifying a model as a path diagram allows one to trace paths showing direct, indirect, and total effects. In the path diagram in Figure 2.2, Panel A, for instance, there are three direct effects. Two coefficients show relationships from exogenous variables to endogenous variables, γ_{11} and γ_{21}. One coefficient depicts a path from one endogenous variable to the other endogenous variable, β_{21}. In this model, there is an indirect effect of x_1 on y_2 that works through y_1. The indirect effect is $\gamma_{11}\beta_{21}$. The total effect of x_1 on y_2 summarizes direct and indirect effects, $\gamma_{21} + \gamma_{11}\beta_{21}$. Chapter 6 will describe calculating and testing indirect and total effects in more detail.

Structural Versus Reduced-Form Equations

Thus far, we have focused our attention on the structural equations of the model. Structural equations represent the theoretical model, showing direct relationships between the variables. Structural parameters summarize the direct, "causal" links between variables. Equations 2.19 and 2.20 are examples of the structural equations of a model. Models can also be written in "reduced form," as reduced-form equations.

Reduced-form equations express the endogenous variables solely as a function of the exogenous variables. That is, only exogenous variables appear on the right-hand side (RHS) of the equations. In a model with a reciprocal path, creating the reduced-form equations entails collecting endogenous variables on the left-hand side. In any model, there are the same number of structural and reduced-form equations.

Returning to the nonrecursive model presented in Figure 2.3, Panel A, the structural equations are as follows:

$$y_1 = \beta_{12}y_2 + \gamma_{11}x_1 + \gamma_{12}x_2 + \zeta_1 \tag{2.22}$$

$$y_2 = \beta_{21}y_1 + \gamma_{22}x_2 + \gamma_{23}x_3 + \zeta_2 \tag{2.23}$$

To determine the reduced-form equation for the y_1 equation, substitute for y_2, yielding

$$y_1 = \beta_{12}(\beta_{21}y_1 + \gamma_{22}x_2 + \gamma_{23}x_3 + \zeta_2) + \gamma_{11}x_1 + \gamma_{12}x_2 + \zeta_1$$

Next, multiply, gather, and rearrange terms.

$$y_1 = \frac{1}{1 - \beta_{12}\beta_{21}}(\gamma_{11}x_1 + \beta_{12}\gamma_{22}x_2 + \gamma_{12}x_2 + \beta_{12}\gamma_{23}x_3 + \beta_{12}\zeta_2 + \zeta_1)$$

The reduced-form equation is therefore

$$y_1 = \Pi_{11}x_1 + \Pi_{12}x_2 + \Pi_{13}x_3 + \zeta_1^* \tag{2.24}$$

where

$$\Pi_{11} = \frac{\gamma_{11}}{1 - \beta_{12}\beta_{21}}$$

$$\Pi_{12} = \frac{\beta_{21}\gamma_{22} + \gamma_{12}}{1 - \beta_{12}\beta_{21}}$$

$$\Pi_{13} = \frac{\beta_{12}\gamma_{23}}{1 - \beta_{12}\beta_{21}}$$

$$\zeta_1^* = \zeta_1 + \beta_{12}\zeta_2$$

The y_2 equation is similar:

$$y_2 = \Pi_{21}x_1 + \Pi_{22}x_2 + \Pi_{23}x_3 + \zeta_2^* \tag{2.25}$$

where

$$\Pi_{21} = \frac{\beta_{21}\gamma_{11}}{1 - \beta_{12}\beta_{21}}$$
$$\Pi_{22} = \frac{\beta_{21}\gamma_{12} + \gamma_{22}}{1 - \beta_{12}\beta_{21}}$$
$$\Pi_{23} = \frac{\gamma_{23}}{1 - \beta_{12}\beta_{21}}$$
$$\zeta_2^* = \beta_{21}\zeta_1 + \zeta_2$$

Reduced-form equations provide information about the total effects of exogenous variables on endogenous variables in a model. In Equation 2.25, Π_{21} is the total effect of x_1 on y_2, which includes the indirect effect of x_1 on y_2 that works through y_1 and includes the reciprocal relationship between y_1 and y_2. Chapter 6 discusses the multiplier, $1/(1 - \beta_{12}\beta_{21})$. The reduced-form equations are central to understanding assessment of simultaneous equation models, as we discuss in Chapter 5.

Instrumental Variables

Identifying and estimating nonrecursive models require understanding instrumental variables. Instrumental variable (IV) estimation was developed for situations where the regressor is correlated with the error term such as occurs in nonrecursive models. In such situations, the regressor is sometimes called "troublesome" or "problematic." Although instrumental variables are frequently treated as a technical solution to an identification or estimation problem—and indeed, this is an important function they serve—we discuss them in this chapter because our view is that they should fundamentally arise from serious theoretical consideration. Furthermore, although instrumental variables are most frequently associated with the literature on limited-information estimators, we emphasize throughout this monograph that careful selection of instrumental variables is necessary regardless of which estimator the researcher employs.

With a problematic regressor, a researcher must find an instrumental variable, which we will call z_1, that is (1) uncorrelated with the disturbance,

$$\text{COV}(z_1, \zeta_1) = 0$$

but (2) correlated with the variable for which it is an instrument,

$$\mathrm{COV}(z_1, x_1) \neq 0$$

In this monograph, we focus on the need for instrumental variables to address a correlation between the error and a regressor due to the regressor's reciprocal relationship with another variable (discussed in Chapter 4).[7] Instrumental variables are also used to correct for a correlation between a regressor and the error due to other causes, including an omitted variable that correlates with the regressor and affects the dependent variable, or measurement error in the regressor.[8]

Example of Instrumental Variables:
Voluntary Associations and Generalized Trust

We demonstrate the use of instrumental variables in a nonrecursive model with an empirical example. This example, from the political science and sociology literatures, will be used throughout the monograph. The theoretical question focuses on the likely interdependent relationship between voluntary association membership and generalized trust. Research on social capital posits the importance of both voluntary associations and trust to the well-being of society (Fukuyama, 1995; Paxton, 2002; Putnam, 1993). But the relationship between the two is likely reciprocal (Brehm & Rahn, 1997; Claibourn & Martin, 2000; Shah, 1998). For instance, scholars have argued that participation in voluntary associations can bring about greater general trust in others due to repeated social interactions, norms of cooperation, and reputation effects (Paxton, 2007). Conversely, those who are more trusting may feel more comfortable interacting with others in an association and therefore may be more likely to join.

We can measure both voluntary associations and trust using data from the 1993 and 1994 waves of the General Social Survey (GSS). In the GSS, respondents report whether they belong to any of 16 types of voluntary organizations, including service, political, youth, and church groups. We computed the number of types of organizations to which a respondent belonged. Generalized trust is measured by three variables: the extent to which the respondent feels people in general (1) are fair, (2) are helpful, and

---

[7]In the context of simultaneous equation models, the problematic regressor will typically be indicated by a y—as an endogenous variable in the system of equations.

[8]Both the well-known omitted variable problem and measurement error are issues that confront all analytic techniques using observational data, including ordinary least squares, logistic regression, and so on, and are not issues unique to simultaneous equations models.

(3) can be trusted.[9] For our simple example, we created a single factor score estimate that is the combination of the three indicators of trust.[10]

To keep the example simple, we include one predictor that influences both endogenous variables: the level of education of the respondent in years. Theory and prior research suggest that more educated respondents should belong to more voluntary associations and also have greater generalized trust. To identify and estimate this model, we also need at least one instrumental variable for each endogenous variable. This must be a variable that does not have a direct relationship with the outcome variable in a given equation, or any omitted variables that influence the outcome variable. In a path diagram, an instrumental variable will appear as a variable predicting one endogenous variable but not the other.

For voluntary association membership, one such possible instrumental variable is the presence of young children less than 6 years of age. Although there is little reason to expect that the presence of young children will affect one's trust in others, they likely influence the amount of time available for association participation. For trust, a possible instrument is a measure of whether the respondent has been burglarized in the past year. Although there seems little reason for such an event to affect one's memberships, experiencing a burglary should affect one's trust in others.

Figure 2.4 displays this model in a path diagram. The variables y_1, voluntary association memberships, and y_2, generalized trust, are in a reciprocal relationship; x_1, children less than 6 years of age, influences voluntary association memberships but not trust and therefore serves as an instrumental variable in the model; x_2, education, influences both voluntary associations and generalized trust; and x_3, burglary, is hypothesized to influence generalized trust but not membership in voluntary associations. The variable x_3 therefore serves as an instrument for generalized trust in the model displayed in Figure 2.4. The reasoning provided for the instrumental variables in this model is at this point entirely theoretical. Tests described in Chapter 5 can help a researcher determine whether an instrumental variable is valid.

It is also helpful to have an overidentified model. To overidentify the model, we modify the exactly identified model by adding a second excluded instrumental variable to each equation (see Figure 2.5). In the

---

[9]There are three questions regarding trust, helpfulness, and fairness: "Generally speaking, would you say that most people can be trusted or that you can't be too careful in life?" "Would you say that most of the time people try to be helpful, or that they are mostly just looking out for themselves?" "Do you think most people would try to take advantage of you if they got a chance, or would they try to be fair?"

[10]One could instead model trust as a latent variable with three indicators (Paxton, 1999), which would make the resulting example a general structural equation model (Bollen, 1989b).

Figure 2.4 Path Model of Voluntary Associations and Generalized Trust

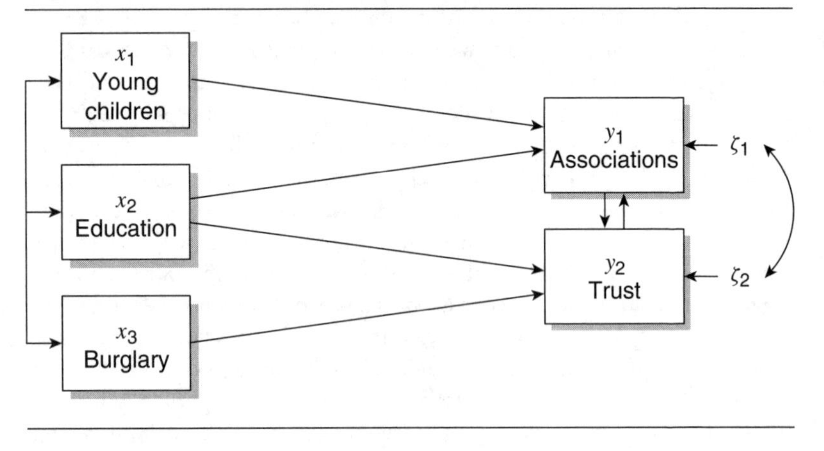

Figure 2.5 Overidentified Model of Voluntary Associations and Generalized Trust

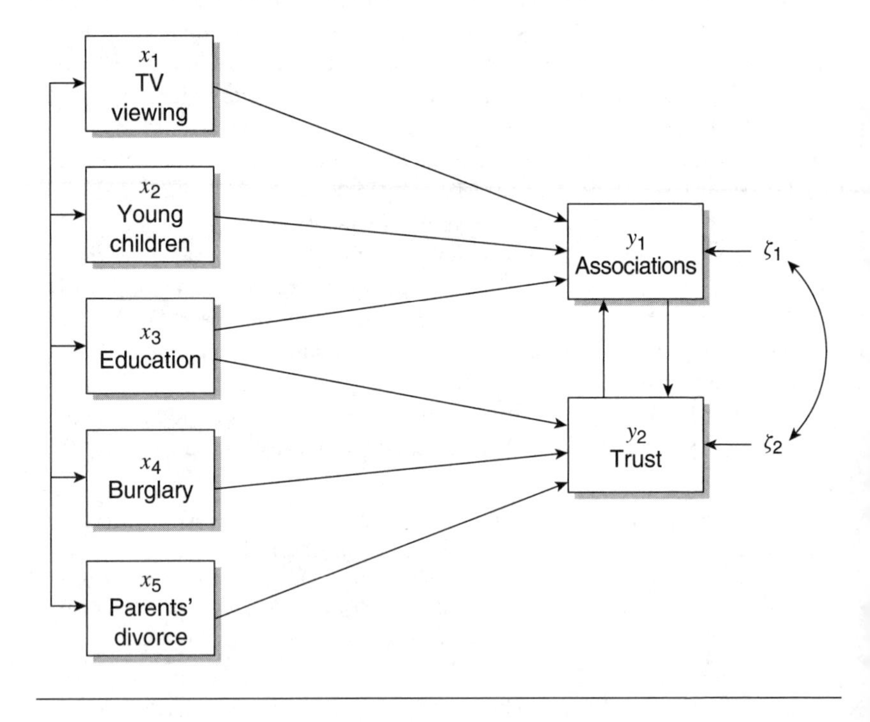

overidentified model, hours of TV viewing influences association participation, and having experienced the divorce of one's parents influences trust.

Examples of Instrumental Variables in Published Research

Finding appropriate instrumental variables is a challenging task and should not be taken lightly. But although challenging, the task is not impossible. In this section, we provide additional examples of instrumental variables used to identify and estimate models in published research.

- Kritzer (1984) used survey data from husbands and wives to test the likely reciprocal effect of political party identification between spouses. That is, husbands' political party identification may influence wives' party identification, and vice versa. Kritzer used the party identification of parents as instruments. His logic was that although we might expect the party identification of a spouse's, say the wife's, parents to affect her party identification, there is little reason to suppose that the wife's parents' party identification would affect her husband's party identification (after accounting for her own party identification). Parents' party identification acts as an instrumental variable in this model.
- Levitt (1996) wanted to investigate the effect of prison incarceration on crime rates. Certainly, increasing the prison population could lead to decreased crime rates. But increased crime rates could also increase incarceration, creating a reciprocal relationship and a problematic regressor. Levitt used litigation related to prison overcrowding as an instrument for prison incarceration. He argued that prison litigation could cause prison populations to decrease but would not directly affect crime rates. In a similar manner, Hoxby (1996) used the passage of laws allowing union activity as an instrument for teachers' unionization in its effect on student outcomes.
- Some studies use attitudinal measures as instruments. A study of the reciprocal relationship between females' labor force participation and fertility expectations used a measure of attitude toward the workplace to predict labor force participation (assuming that it does not directly affect a respondent's fertility expectations) (Waite & Stolzenberg, 1976). This same study also posited that one's ideal family size will affect fertility expectations but will have no direct effect on actual labor force participation.

- Sadler and Woody (2003) theorize that individuals in interaction are influenced by their interaction partner's level of dominance (a reciprocal relationship between two individuals). To create instruments, Sadler and Woody rely on other theories suggesting that individuals bring their own interpersonal style to an interaction as well. Thus, both individuals will bring their own "trait dominances" to the interaction, and trait dominance acts as an instrumental variable in the model. That is, an individual's trait dominance will influence his or her own interactional dominance. But it should not influence his or her interaction partner's interactional dominance directly.
- A study estimating the reciprocal relationship between crime and residential mobility posited population size as an instrument for crime (Liska & Bellair, 1995). Although population size may influence crime, there is little reason to suppose it affects residential mobility. The authors also hypothesized that government revenue per capita—as a proxy for tax burden—may affect residential mobility without affecting crime (making tax burden an instrument for residential mobility). Note that this latter specification assumes that social services do not affect the level of crime, which may or may not be theoretically justified. Statistical tests described in Chapter 5 can help a researcher further evaluate potential instrumental variables.
- Barro and McCleary (2003) investigate whether the religiosity of a population influences a country's economic growth. But there could be a return effect since economic development may cause individuals to become less religious (the secularization hypothesis). As one instrument for religion, Barro and McCleary use the establishment of a state religion. Establishment of a state religion should arguably influence religiosity in a country, but, as state religions were often declared centuries ago, not economic growth in the present period. See Young (2009) for an outstanding discussion and assessment of state religion as an instrument for religiosity.
- Ansolabehere and Jones (2010) ask whether a constituent's perceived agreement with their congressperson on policy issues increases their approval of that congressperson. But it might be that constituents who approve of their congressperson tend to assume they agree on questions of policy. Ansolabehere and Jones had information on *actual* roll call voting by congresspeople and could use that as an instrument for constituents' perceived roll call voting. Reasonably, actual roll call votes can only influence constituent approval through constituent perceptions of those votes.

These are just a few examples of the variables researchers have used as instrumental variables when modeling nonrecursive models. Careful theory and some creativity can produce useful instrumental variables. Indeed, solid reasoning and clever research designs can take precedence over many of the statistical features of these models. In research using instrumental variables to address problematic regressors of a variety of types, researchers have used physical features as instrumental variables (e.g., the number of rivers in a metropolitan area as an exogenous source of the number of political units and subsequent segregation; Cutler & Glaeser, 1997); a lagged version of the variable of interest (assuming that there are no additional effects from earlier time points; Markowitz, Bellair, Liska, & Liu, 2001); and "simulated instruments" (Hoxby, 2001). For other examples, see Murray (2006b, chap. 13).

CHAPTER 3. IDENTIFICATION

The simultaneous estimation of multiple equations introduces a number of complications. One complication is the need to establish identification. Identification is the process whereby researchers theoretically establish that a unique solution exists for each parameter of the model. A simple example from Bollen (1989b) can help illustrate the issue. Consider the following equation:

$$\text{VAR}(y) = \theta_1 + \theta_2 \quad (3.1)$$

Do unique values for θ_1 and θ_2 follow from this equation? No. If $\text{VAR}(y) = 10$, then ($\theta_1 = 2$, $\theta_2 = 8$) and ($\theta_1 = -1$, $\theta_2 = 11$) are both valid solutions. In fact, there are an infinite number of viable solutions for θ_1 and θ_2. With two unknowns and one equation, θ_1 and θ_2 are not identified.

Adding a second equation

$$\theta_1 = \theta_2 \quad (3.2)$$

produces identified parameters. If $\text{VAR}(y) = 10$, then $\theta_1 = \theta_2 = 5$.

From this example, it should be clear that identification is a mathematical problem, not a statistical problem. Furthermore, the example highlights that identification is a property of a model and not of the data. Collecting more data will not change the identification status of a model.

Although identification is often described in terms of equations or entire models, it is at base a property of parameters. Unidentified parameters cannot be consistently estimated, and a single unidentified parameter will underidentify an entire model.[1] Establishing identification is required before estimation so that a researcher can be confident that he or she will obtain unique estimates of the coefficients.

Known and Unknown Parameters

We can distinguish between two types of parameters: "known-to-be-identified" parameters and "unknown" parameters. Initially, known parameters

[1]Model underidentification does not imply that all parameters in the model are unidentified. In fact, underidentified models can contain within them equations that are identified. In rare cases, a researcher may want to estimate the identified parts of an overall underidentified model. We recommend caution when attempting to work with underidentified models.

are the population moments (variances and covariances) of the observed variables. Unknown parameters appear in $\boldsymbol{\theta}$, which includes all paths, all variances of the exogenous variables, all covariances between the exogenous variables, all variances of the disturbances, and all covariances between disturbances. In other words, it includes all parameters to be estimated in the model appearing in $\mathbf{B}$, $\boldsymbol{\Gamma}$, $\boldsymbol{\Phi}$, and $\boldsymbol{\Psi}$. The goal in identification is to show that all unknown parameters are unique functions of known-to-be-identified parameters.

At the start of the identification process, the number of known parameters is equal to $(p + q)(p + q + 1)/2$, which is the number of nonredundant elements of the variance/covariance matrix of observations. For instance, if you had three variables (two xs and one y), you will have six nonredundant pieces of information from the covariance matrix:

$$\begin{bmatrix} \mathrm{VAR}(y_1) & & \\ \mathrm{COV}(x_1, y_1) & \mathrm{VAR}(x_1) & \\ \mathrm{COV}(x_2, y_1) & \mathrm{COV}(x_2, x_1) & \mathrm{VAR}(x_2) \end{bmatrix} \tag{3.3}$$

These are known to be identified. In simple models, if each parameter in the model can be written as a unique function of these variances and covariances, then the model is identified. For example, consider a simple regression (see Equation 2.13) with three variances and covariances:

$$\begin{aligned} \mathrm{VAR}(x_1) &= \phi_{11} \\ \mathrm{COV}(x_1, y_1) &= \gamma_{11}\phi_{11} \\ \mathrm{VAR}(y_1) &= \gamma_{11}^2\phi_{11} + \psi_{11} \end{aligned}$$

Solving for ϕ_{11} is straightforward as it is equal to $\mathrm{VAR}(x_1)$. Next, γ_{11} can be solved:

$$\begin{aligned} \gamma_{11} &= \mathrm{COV}(x_1, y_1)/\phi_{11} \\ &= \mathrm{COV}(x_1, y_1)/\mathrm{VAR}(x_1) \end{aligned}$$

Note that once ϕ_{11} is shown to be known, it can be used to identify other unknown parameters. With both γ_{11} and ϕ_{11} shown to be identified, identifying ψ_{11} is now straightforward:

$$\psi_{11} = \mathrm{VAR}(y_1) - \gamma_{11}^2\phi_{11}$$

Of course, some models are too complicated to be identified algebraically in this way. Still, the distinction between known and unknown parameters is useful for understanding other ways to identify models. For example, building on the mathematical principle of equations and unknowns, it is clear that, at a minimum, we must have at least as many variances and covariances as the number of parameters to identify a model.

Restrictions

Identification requires that some restrictions be placed on model parameters. Typically these restrictions take the form of a zero restriction, setting a parameter to zero, although other types of restrictions are possible. Two types of restrictions are commonly assumed but not often recognized by researchers. The first such implicit restriction is that the main diagonal of **B** is fixed to zero. The theoretical implication of this restriction is that a variable cannot have an instantaneous effect on itself. The second implicit restriction is that the parameter matrix for the errors in the equations is fixed to the identity matrix (each coefficient is set to 1). The rationale for this restriction is that, as unobserved variables, each error must be scaled (see Bollen, 1989b, p. 91).

Beyond these implicit restrictions, we use theory to place restrictions on some parameters of the model. For example, we may set a path to zero, as in Figure 2.4, where the path from x_1 to y_2 is set to zero. Or we may set the correlation between two errors to be zero, as in Figure 2.1.

The focus of restrictions in this text is largely on exclusion restrictions, where parameters are fixed to zero. Restricting certain paths to zero is a key feature of instrumental variables in simultaneous equation models. Other restrictions are possible, such as specifying that two parameters are equivalent, or specifying that one parameter is a function of other parameters.

Three Types of Models

We distinguish three types of models: underidentified models, just-identified models, and overidentified models. Underidentified models are those in which there are too few known parameters to identify all the unknown parameters. In underidentified models, at least one parameter is underidentified. As there is no way to meaningfully estimate an underidentified parameter, a researcher should not proceed to estimation in the presence of an underidentified model. Instead, a researcher must either impose further restrictions or add information to identify the model. For example, consider Figure 3.1, which is currently underidentified.

There are several ways whereby we could identify this model. For example, if we remove the path from x_1 to y_2 from this model, it would be identified. Or if we remove the paths from x_2 or x_3 to y_2, the model would be identified. Alternatively, we could add a variable, x_5, that influences only y_1. That would also identify the model by adding information. As will be discussed in detail below, such restrictions and additions must always be theoretically appropriate.

Models that meet the minimum conditions for uniquely solving for each parameter are said to be "just-identified." Just-identified models are also known as exactly identified models or as saturated models. With just-identified models, the model contains an equal number of known and unknown parameters and the model is identified. Remembering that the known parameters are contained in the variances and covariances of the observed variables, this means that in just-identified models we have the same number of variances and covariances as parameters. Furthermore, each parameter is identified, and none are overidentified.

With an overidentified model, we have more information than we need to solve for the parameters. Overidentified models have fewer unknowns than knowns such that there is more than one way to solve for a parameter, and all parameters have at least one unique solution. In overidentified

Figure 3.1 Underidentified Model

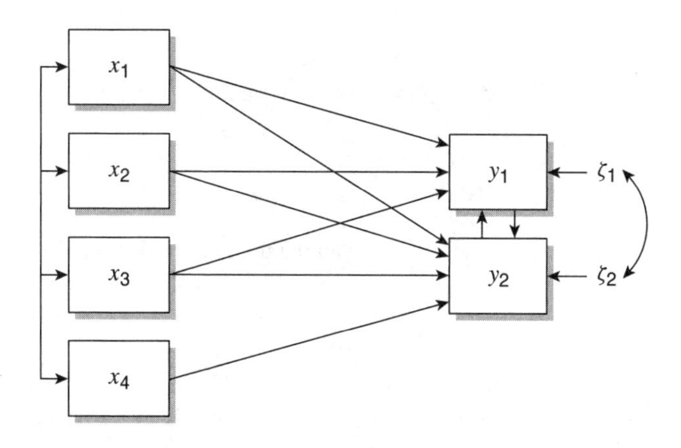

models at least one parameter can be expressed as two or more functions of the variances and covariances of the model.

For example, consider Figure 3.2.

Figure 3.2 Simple Overidentified Model

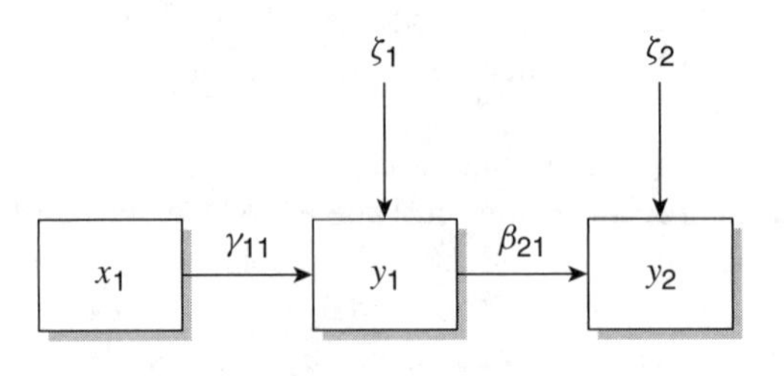

There is more than one solution for β_{21}. Using covariance algebra, we can determine the observed and implied covariance matrices, $\mathbf{\Sigma} = \mathbf{\Sigma(\theta)}$:

$$\begin{bmatrix} \mathrm{VAR}(y_1) & & \\ \mathrm{COV}(y_2, y_1) & \mathrm{VAR}(y_2) & \\ \mathrm{COV}(x_1, y_1) & \mathrm{COV}(x_1, y_2) & \mathrm{VAR}(x_1) \end{bmatrix} = \begin{bmatrix} \gamma_{11}^2\phi_{11} + \psi_{11} & & \\ \beta_{21}\gamma_{11}^2\phi_{11} + \beta_{21}\psi_{11} & \beta_{21}^2\gamma_{11}^2\phi_{11} + \beta_{21}^2\psi_{11} + \psi_{22} & \\ \gamma_{11}\phi_{11} & \beta_{21}\gamma_{11}\phi_{11} & \phi_{11} \end{bmatrix}$$

We can use the resulting equations to solve for β_{21}. For example, we can use the model-implied values for $\mathrm{COV}(x_1, y_2)$ and $\mathrm{COV}(x_1, y_1)$:

$$\frac{\mathrm{COV}(x_1, y_2)}{\mathrm{COV}(x_1, y_1)} = \frac{\beta_{21}\gamma_{11}\phi_{11}}{\gamma_{11}\phi_{11}}$$

Canceling out the $\gamma_{11}\phi_{11}$ term in the numerator and the denominator yields

$$\beta_{21} = \frac{\mathrm{COV}(x_1, y_2)}{\mathrm{COV}(x_1, y_1)}$$

So β_{21} is now known to be identified. But there is an overidentifying restriction in our model, as β_{21} can be written as a second function of the variances and covariances of the observed variables. We can use the model-implied value for COV(y_2, y_1) and the model-implied value for VAR(y_1) to yield

$$\frac{\text{COV}(y_2, y_1)}{\text{VAR}(y_1)} = \frac{\beta_{21}(\gamma_{11}^2\phi_{11} + \psi_{11})}{\gamma_{11}^2\phi_{11} + \psi_{11}}$$

We can cancel out the same term in the numerator and denominator, leaving

$$\beta_{21} = \frac{\text{COV}(y_2, y_1)}{\text{VAR}(y_1)}$$

Overidentified models are useful because they create testable restrictions. In this example, if each of these functions of covariances equal the parameter, they must also equal each other:

$$\frac{\text{COV}(y_2, y_1)}{\text{VAR}(y_1)} = \frac{\text{COV}(x_1, y_2)}{\text{COV}(x_1, y_1)} \tag{3.4}$$

This is a constraint implied by the model, which we call an overidentifying restriction. Since the restriction is implied by our model, but does not necessarily occur in the data, it is testable. That is, our model implies that the equality in Equation 3.4 holds in the population. We can test whether it holds in our sample within sampling error.

In sum, overidentification produces certain restrictions that can be tested, lending insight into whether our model is a good fit to the data. Understanding overidentifying restrictions helps us understand the fit statistics to be discussed in Chapter 5.

Rules of Identification

Rules exist to help with identification. Rules can be either model based or equation based. Model-based identification rules identify particular types of models as a class. Equation-based identification rules identify models equation by equation. When using an equation-based identification rule, if all equations in the model are shown to be identified, then the model is said to be identified.

Model-Based Identification Rules

t *Rule*

This necessary but not sufficient rule of identification builds on the equations and unknowns problem. The rule states that the number of known variances and covariances of the observed variables must be equal to or greater than the number of unknown model parameters:

$$t \leq \left(\frac{1}{2}\right)(p + q)(p + q + 1) \tag{3.5}$$

Remember that $p + q$ totals the number of observed variables (endogenous and exogenous) in the model. t is the number of parameters to be estimated in the model, and (1/2) $(p + q)(p + q + 1)$ provides the number of nonredundant elements in the covariance matrix of the observed variables (elements in the main diagonal and below).

As a necessary condition, the t rule provides a quick way to establish that some models are *not* identified. But the t rule is not sufficient to identify a model. It is not merely having enough known parameters relative to unknown parameters that matters for identification; the configuration of the unknowns in the model matters as well.

The Recursive Rule

The recursive rule is straightforward: If your model is recursive, it is identified. The recursive rule is sufficient but not necessary to identify a model. Recursive models do not exhibit reciprocal causation or feedback loops, and their disturbances are not correlated. More formally, a lack of reciprocal paths means that $\mathbf{B}$ can be written as lower triangular, with all the elements falling below the diagonal. Furthermore, the $\boldsymbol{\Psi}$ matrix is diagonal, with elements appearing only down the diagonal. Both conditions must be met for a model to be recursive. As an example, consider Figure 2.1, which is a recursive model. Direction of influence in this model is running in only one direction, without reciprocal paths or feedback loops. Furthermore, there are no correlations between the errors in the equations. Thus, the model is recursive and therefore identified. A researcher working with this model would be able to move to the next step in modeling: estimation.

The Null Beta Rule

Models in which no endogenous variable affects another have a $\mathbf{B}$ matrix equal to zero and are identified. The null beta rule is sufficient but not necessary to identify a model. A special class of models, called seemingly unrelated regression, or SUR models, are identified under the null beta rule. Seemingly unrelated regression models (Zellner, 1962) refer to multiple equations that, on the surface, appear to be unrelated to one another, but whose errors are correlated across equations. As an example, consider Figure 3.3.

Figure 3.3 Example of a Seemingly Unrelated Regression Model

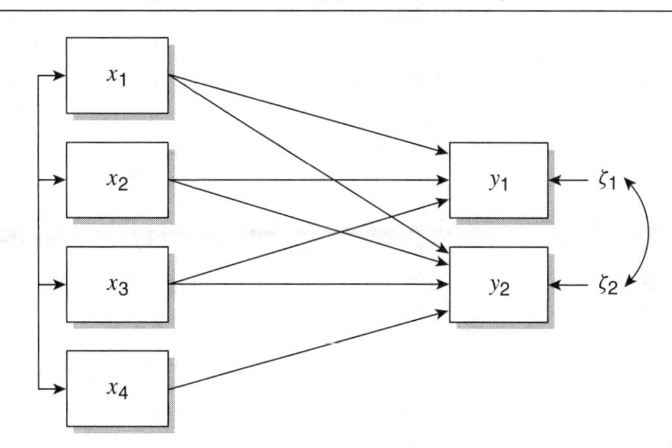

In Figure 3.3, neither endogenous variable affects the other, but the equations are connected through the relationship between ζ_1 and ζ_2. A model of this type is identified through the null beta rule.

Both the recursive rule and the null beta rule are restriction-based identification rules. They determine whether the number and configuration of model parameters are sufficient to identify a model. Unfortunately, nonrecursive models do not satisfy the restriction conditions for either rule. Nonrecursive models require the identification of models on an equation-by-equation basis, as outlined in the next section.

Equation-Based Identification Rules

The Order Condition

There are two equivalent ways to state the order condition:

1. In a model of p simultaneous equations, an equation is identified if it excludes at least $p - 1$ variables (endogenous or exogenous) that appear elsewhere in the model.

2. In a model of p simultaneous equations, an equation is identified if the number of excluded exogenous variables is equal to or greater than the number of endogenous variables in that equation minus one.

In a model, p references the number of endogenous variables. Both ways to state the order condition appear in various textbooks, and the researcher can use whichever is most intuitive. Following the first statement, for a system of p endogenous variables it follows that there are p equations. An equation is identified if $p - 1$ variables are left out of the equation. The excluded variables can be endogenous or exogenous. The order condition is a simple counting rule—does the model restrict a sufficient number of paths, producing excluded variables, for identification?

If the $\mathbf{\Psi}$ matrix contains no restrictions such that all errors in the equation are completely intercorrelated, then the order condition is necessary but not sufficient to identify a model. Certain models will pass the order condition and not be identified. Like the t rule, therefore, the order condition can help researchers rule out underidentified models with full $\mathbf{\Psi}$ matrices.

If the $\mathbf{\Psi}$ matrix does contain restrictions, then the order condition is not necessary. This point is not emphasized in most econometric texts, where it is assumed that $\mathbf{\Psi}$ is full. But many models do imply restrictions on the $\mathbf{\Psi}$ matrix, and these constraints can help identify a model. In short, models that do not have full $\mathbf{\Psi}$ matrices can fail the order condition and still be identified.

An example, illustrated in Figure 3.4, helps illustrate the order condition:

$$y_1 = \beta_{12}y_2 + \gamma_{11}x_1 + \gamma_{12}x_2 + \zeta_1 \tag{3.6}$$

$$y_2 = \beta_{21}y_1 + \gamma_{22}x_2 + \gamma_{23}x_3 + \zeta_2 \tag{3.7}$$

$$y_3 = \beta_{31}y_1 + \beta_{32}y_2 + \zeta_3 \tag{3.8}$$

Figure 3.4 Nonrecursive Path Model

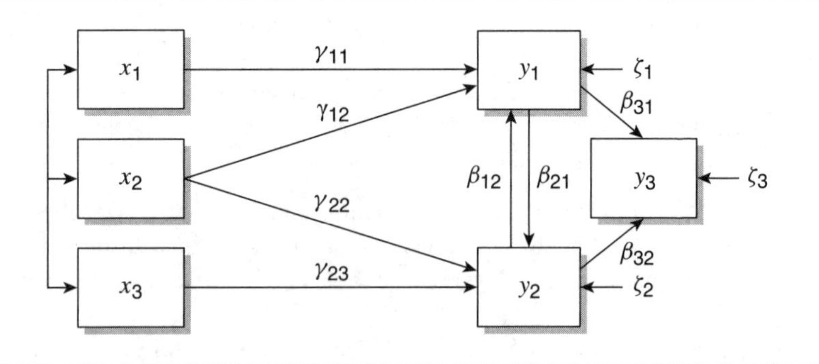

The model needs $p - 1$ omitted variables from each equation, which is two omitted variables in this three-equation system.

Equation 3.6 omits x_3 and y_3 and meets the order condition.

Equation 3.7 omits x_1 and y_3 and meets the order condition.

Equation 3.8 omits x_1, x_2, and x_3 and exceeds the order condition.

Note that the order condition is a necessary but not sufficient condition, so we cannot say for certain that these equations are identified.

For complicated models, creating a matrix of parameters can aid with counting. To create the matrix of coefficients, algebraically move all variables and parameters to the left-hand side of the equation, leaving only the disturbance on the right-hand side.

$$y_1 - \beta_{12}y_2 - \gamma_{11}x_1 - \gamma_{12}x_2 = \zeta_1 \quad (3.9)$$

$$y_2 - \beta_{21}y_1 - \gamma_{22}x_2 - \gamma_{23}x_3 = \zeta_2 \quad (3.10)$$

$$y_3 - \beta_{31}y_1 - \beta_{32}y_2 = \zeta_3 \quad (3.11)$$

Reorder variables appropriately and include one and zero parameters.

$$1y_1 - \beta_{12}y_2 - 0y_3 - \gamma_{11}x_1 - \gamma_{12}x_2 + 0x_3 = \zeta_1 \quad (3.12)$$

$$-\beta_{21}y_1 + 1y_2 - 0y_3 + 0x_1 - \gamma_{22}x_2 - \gamma_{23}x_3 = \zeta_2 \quad (3.13)$$

$$-\beta_{31}y_1 - \beta_{32}y_2 + 1y_3 + 0x_1 + 0x_2 + 0x_3 = \zeta_3 \quad (3.14)$$

And now create a matrix of the coefficients.

$$\mathbf{M} = \begin{bmatrix} 1 & -\beta_{12} & 0 & -\gamma_{11} & -\gamma_{12} & 0 \\ -\beta_{21} & 1 & 0 & 0 & -\gamma_{22} & -\gamma_{23} \\ -\beta_{31} & -\beta_{32} & 1 & 0 & 0 & 0 \end{bmatrix}$$

Now the straightforward counting rule becomes even more apparent. For each row of the matrix, count the zeros. If a row has at least $(p - 1)$ zeros, then its corresponding equation meets the order condition. If all rows have a sufficient number of zeros, then the model as a whole meets the order condition. Here, each row of the matrix has at least two exclusions, and so the model passes the order condition. An equivalent way to create this matrix is simply to form $[\mathbf{I}-\mathbf{B} | -\boldsymbol{\Gamma}]$.

The Rank Condition

Recall that the order condition is necessary but not sufficient. Models where one equation is a linear combination of another equation could pass the order condition but still be underidentified. That is, to determine identification, we need to establish that one equation is not a linear combination of other equations (Duncan, 1975, p. 28). Is the specification of the model, as outlined in the coefficient matrices, sufficient to distinguish the equations of the model from other possible linear combinations of those equations?

To understand the reasoning behind the rank condition, let us modify the simple two-equation example from the beginning of the chapter (see Asher, 1983, pp. 53–54). Rather than the addition of the $\theta_1 = \theta_2$ equation, we add

$$2\text{VAR}(y) = 2\theta_1 + 2\theta_2 \tag{3.15}$$

The addition of this second equation does not identify the model because the second equation, being simply twice the first, does not add new information to the model. To address this concern, we need a rule that is *necessary* and *sufficient*—the rank condition.

The rank condition can be concisely stated as follows: With p equations and p endogenous variables, an equation is identified if at least one nonzero determinant of order $(p - 1)(p - 1)$ can be constructed from the coefficients of the variables excluded from that equation (but included in other equations).

Like the order condition, the rank condition assumes a full $\boldsymbol{\Psi}$ matrix, where all errors in the equations are completely intercorrelated. If $\boldsymbol{\Psi}$ is not full, then the rank condition is sufficient to identify the model, but not necessary.

To determine the rank of a matrix, a four-step procedure is recommended.

1. Form a matrix of the coefficients. For the example from above, this would be as follows:

$$\mathbf{M} = \begin{bmatrix} 1 & -\beta_{12} & 0 & -\gamma_{11} & -\gamma_{12} & 0 \\ -\beta_{21} & 1 & 0 & 0 & -\gamma_{22} & -\gamma_{23} \\ -\beta_{31} & -\beta_{32} & 1 & 0 & 0 & 0 \end{bmatrix}$$

2. Strike out the row of the equation under consideration. For example, for the first equation we would delete the first row.

$$\mathbf{M}_1 = \begin{bmatrix} 1 & -\beta_{12} & 0 & -\gamma_{11} & -\gamma_{12} & 0 \\ -\beta_{21} & 1 & 0 & 0 & -\gamma_{22} & -\gamma_{23} \\ -\beta_{31} & -\beta_{32} & 1 & 0 & 0 & 0 \end{bmatrix}$$

3. Strike out all columns that do not have zeros in the equation under consideration. For example, for the first equation we would delete the first, second, fourth, and fifth columns.

$$\mathbf{M}_1 = \begin{bmatrix} 1 & \beta_{12} & 0 & \gamma_{11} & \gamma_{12} & 0 \\ -\beta_{21} & 1 & 0 & 0 & -\gamma_{22} & -\gamma_{23} \\ -\beta_{31} & -\beta_{32} & 1 & 0 & 0 & 0 \end{bmatrix}$$

4. Use the remaining values to form a submatrix (e.g., $\mathbf{M}_1$) and determine whether there is a nonzero determinant of order $(p - 1)(p - 1)$. For the first equation, we would be left with the following submatrix:

$$\mathbf{M}_1 = \begin{bmatrix} 0 & -\gamma_{23} \\ 1 & 0 \end{bmatrix}$$

$$\begin{aligned} |\mathbf{M}_1| &= 0 - (-\gamma_{23}) \\ &= \gamma_{23} \\ &\neq 0 \end{aligned}$$

The first equation has a nonzero determinant of order 2×2; therefore Equation 1 is identified by the rank condition. For more information on determinants, see Gill (2006, chap. 4).

To finish identifying the model, we would follow the same steps for Equations 2 and 3. For Equation 2, we have the following:

$$\mathbf{M}_2 = \begin{bmatrix} 0 & -\gamma_{11} \\ 1 & 0 \end{bmatrix}$$

$$\begin{aligned} |\mathbf{M}_2| &= 0 - (-\gamma_{11}) \\ &= \gamma_{11} \\ &\neq 0 \end{aligned}$$

Thus, Equation 2 is identified. For Equation 3,

$$\mathbf{M}_3 = \begin{bmatrix} -\gamma_{11} & -\gamma_{12} & 0 \\ 0 & -\gamma_{22} & -\gamma_{23} \end{bmatrix}$$

$\mathbf{M}_3$ has rank = 2; thus Equation 3 is also identified.

All three equations in this three-equation system have been shown to be identified using the rank condition. Therefore, the model as a whole is identified. For a way to determine the rank condition without use of matrices, readers are encouraged to turn to an earlier volume in this series, Berry (1984).

Block-Recursive Identification

For nonrecursive models without fully correlated errors, the rank condition is no longer necessary. That is, there are some such models that will fail the rank condition but still be identified. For these models, we may gain traction on the issue of identification by breaking the model into groups of equations. Block-recursive models (Fisher, 1961) segregate equations into groups (blocks), where reciprocal relationships, feedback loops, or correlated errors can occur *within* blocks of variables, but relations *between* blocks are recursive. In short, we shift our frame of reference slightly, no longer thinking of the model as single equations. Rather, we think of the model as being subdivided into blocks of equations.

By reformulating a nonrecursive model as a block-recursive model, we gain a number of benefits. First, looking across the blocks, we create a system with desirable properties (here recursivity) that is identified. Second, within each nonrecursive block, identification is often greatly eased. As pointed out by Edward Rigdon (1995), most models have a tendency to reduce to blocks of one or two equations. The full set of possible two-equation blocks falls into a limited set of classes that Rigdon concisely provides as eight special cases. Knowledge of the eight possible two-equation blocks can identify most block-recursive models efficiently and accurately. If we determine that each of the blocks is identified, and we know that the model across blocks is identified (due to the recursive rule), we can conclude that the model as a whole is identified.[2]

Blocking Models

The first step in identifying a model through a block-recursive technique is to segment the model into blocks. Make sure to segment the model as far as possible (Rigdon, 1995). That is, if the equations within a block are recursively related, further segment the block, even to single-equation blocks.

---

[2]It is often also possible to assess the identification of these blocks by using the order or rank conditions described above. The choice is that of the researcher. Our experience is that many researchers find the eight special cases described in this section to be quite intuitive and therefore easy to use.

As an example, consider the (adapted) model (Figure 3.5) presented in Ethington and Wolfle (1986).

Figure 3.5 Ethington and Wolfle (1986) Model

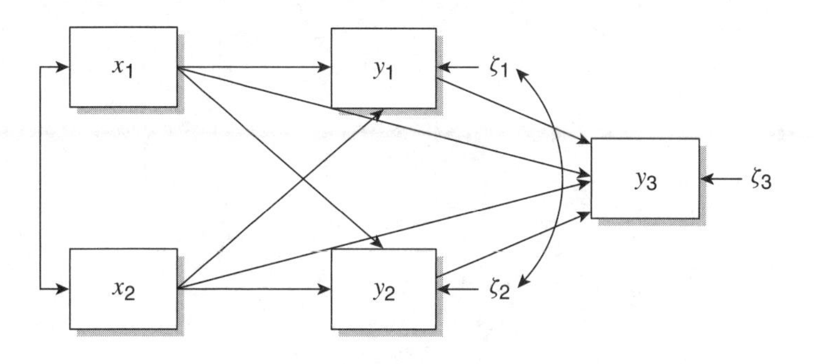

This model can be segmented into two blocks, as in Figure 3.6. The first block is nonrecursive because y_1 and y_2 have correlated errors. The second block has only a single equation and is therefore recursive. Although the model as a whole is nonrecursive, the first block has a recursive effect on the second block.

Figure 3.6 Ethington and Wolfle (1986) Model, Blocked

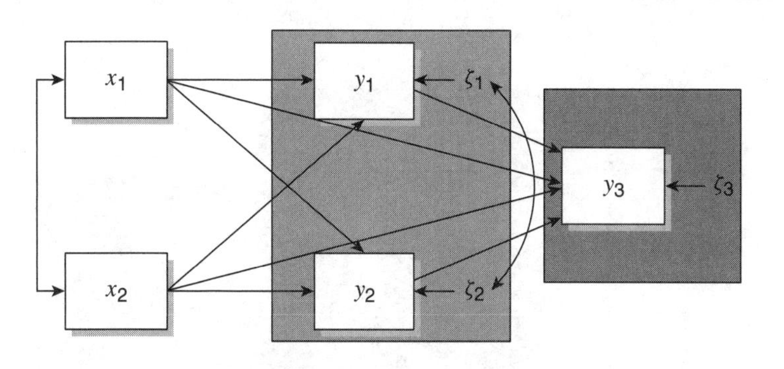

As a second example, consider Figure 3.7, a model presented in Duncan, Featherman, and Duncan (1972).[3]

[3]For pedagogical purposes, the Duncan et al. (1972) model has been modified slightly from the original.

Figure 3.7 Duncan et al. (1972) Model

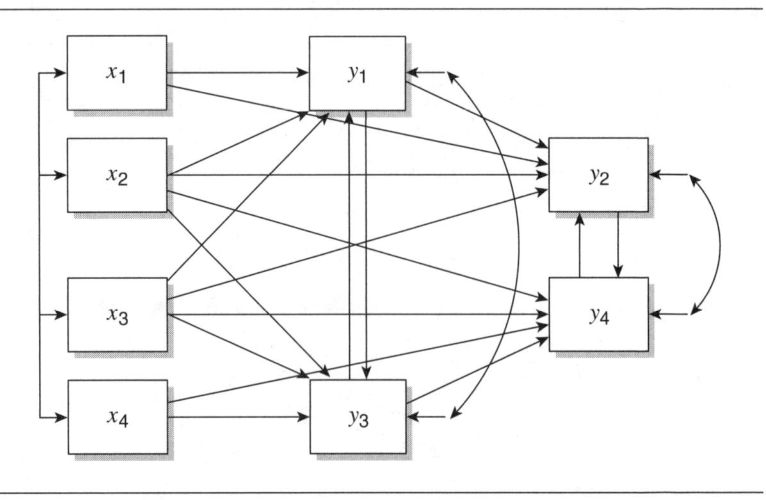

Again, this model has two blocks, as illustrated in Figure 3.8. The first block captures the nonrecursive relationship between y_1 and y_3, while the second captures the nonrecursive relationship between y_2 and y_4. As there are no reciprocal relationships, feedback loops, or correlated errors between the blocks, the system is block recursive.

Figure 3.8 Duncan et al. (1972) Model, Blocked

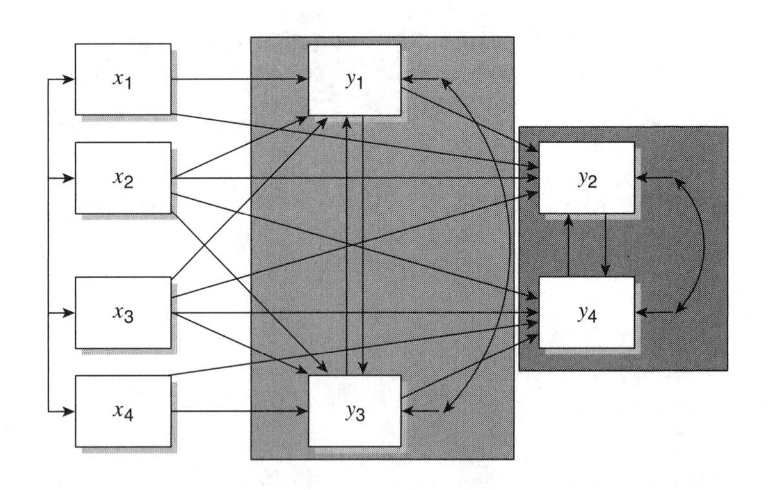

Each of these two examples is recursive across blocks and therefore identified across blocks. What remains to be determined is whether they are identified within the blocks. The benefit to blocking the model in this way is that a researcher can now evaluate identification one block at a time.

Identification of Two-Equation Blocks

If we segment a model into two equation blocks, then there are eight cases we can discuss (Rigdon, 1995). Rigdon estimates that block-recursive models that break into two-equation blocks represent 81% of nonrecursive structural models without full $\mathbf{\Psi}$.

In this section, we present each of these eight possible two-equation blocks. In each diagram, information relevant to identification appears in solid lines, while irrelevant information (for identification) appears in dashed lines. For example, in Case 1 the important pieces of the block are that there are two endogenous variables that have correlated errors. Irrelevant to identification are the number of unique exogenous predictors of each. Omissions are also important in the diagrams. In Case 1, y_1 does not influence y_2, and y_2 does not influence y_1.

Note that the first seven cases are focused on unique predictors, and the last case (Case 8) describes common predictors.

Case 1: Identified

The case of two endogenous variables linked only through their error terms is identified by the null beta rule (Figure 3.9). It does not matter whether either endogenous variable has a unique predictor. Any block in the block-recursive system without **B** parameters linking the endogenous variables is identified.

Figure 3.9 Case 1

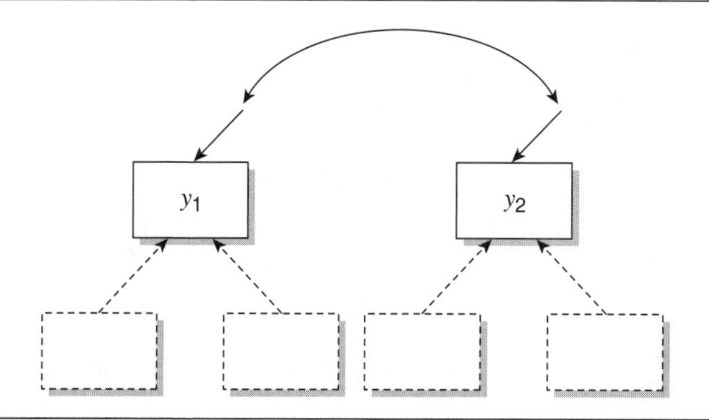

Case 2: Not Identified

In Case 2 blocks, errors are correlated, y_1 influences y_2 (β_{21}), y_1 does not have any unique predictors, and y_2 may or may not have unique predictors. In such a block, there are too few known-to-be-identified parameters to identify the unknown parameters (Figure 3.10).

Figure 3.10 Case 2

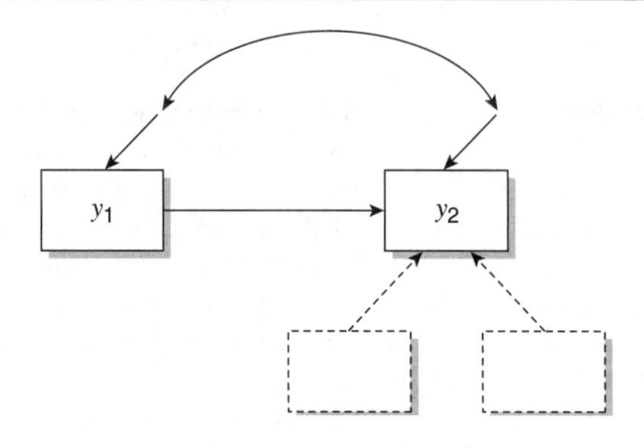

Case 3: Identified

Case 3 resembles Case 2 in that errors are correlated, and y_1 influences y_2. But critically for identification, in Case 3 y_1 has at least one unique predictor. A block with a form like Case 3 will be identified (Figure 3.11).

Figure 3.11 Case 3

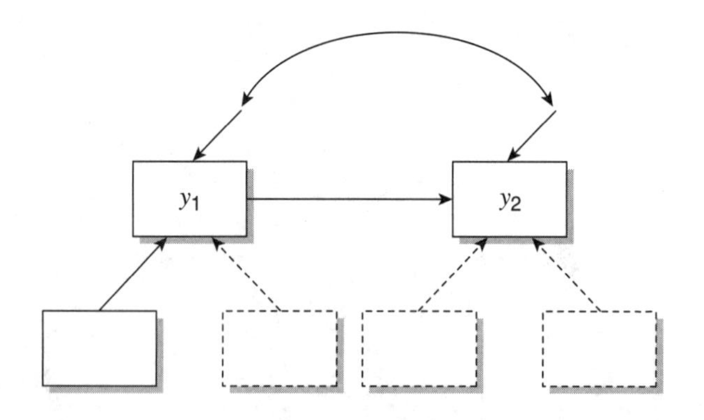

Case 4: Not Identified

In Case 4, y_1 and y_2 are in a reciprocal relationship and neither has a unique predictor. Without unique predictors, whether or not the errors are correlated (as indicated by the dashed line), the block is not identified (Figure 3.12). There are simply too many unknowns relative to knowns to identify the model.

Figure 3.12 Case 4

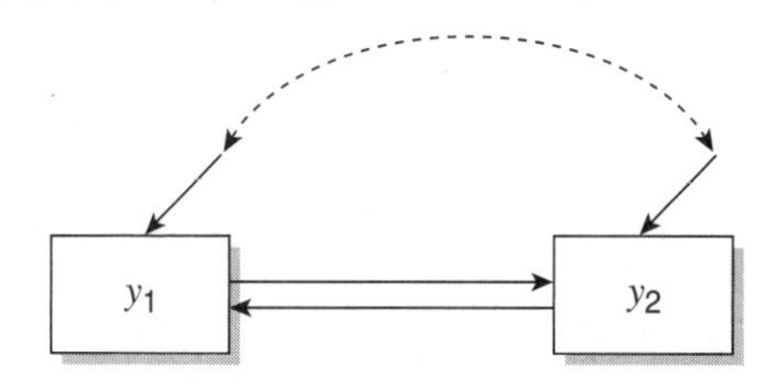

Case 5: Identified

Case 5 demonstrates one way to identify a reciprocal relationship between two variables. If researchers have a unique predictor for either y_1 or y_2 and can make the assumption that there is no correlation between the error terms, the block is identified (Figure 3.13). The omission of a correlation between ζ_1 and ζ_2 is important here. Identifying this block depends on understanding that the order and rank conditions are sufficient but not necessary when there are restrictions in the $\mathbf{\Psi}$ matrix. Here, the zero restriction between the errors for y_1 and y_2 helps us identify the block, even though the block fails the rank and order conditions (see Rigdon, 1995, pp. 380–382).

Figure 3.13 Case 5

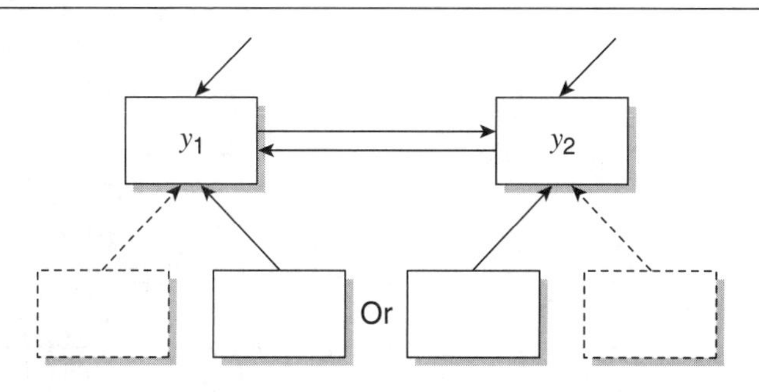

Case 6: Not Identified

It follows from our discussion of Case 5 that Case 6 is not identified (Figure 3.14). With a reciprocal relationship, a unique predictor for only one of the two endogenous variables, and correlated errors, this block fails the order and rank conditions.

Figure 3.14 Case 6

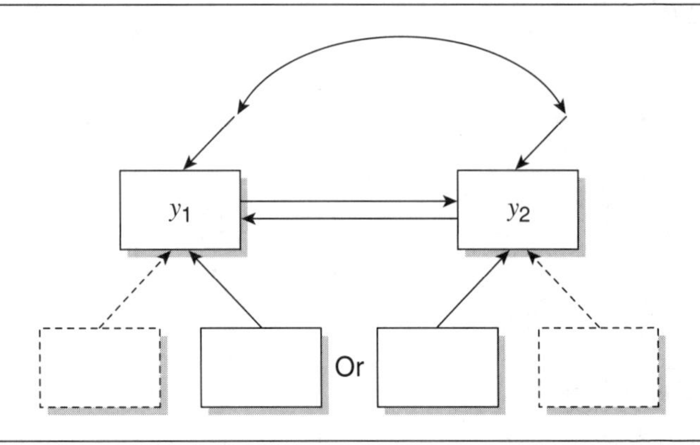

Case 7: Identified

Case 7 illustrates the well-known situation where each endogenous variable in a reciprocal relationship has an associated unique predictor. Regardless whether the errors are correlated, Case 7 meets the order and rank conditions and is therefore identified (Figure 3.15).[4]

Figure 3.15 Case 7

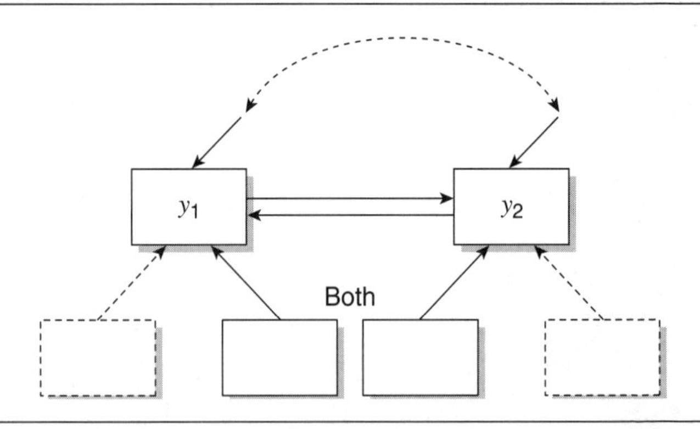

[4]Feedback loops that involve more than two variables will not fall into one of these eight cases. Similar principles apply, however. For example, if a unique instrument is associated with each variable in a feedback loop, then the model is identified (Heise, 1975, p. 180).

Case 8: Reclassify

The purpose of Case 8 is to address shared predictors. Note that in Figure 3.16 everything about the model is irrelevant (dashed) except for a shared predictor. All the preceding cases deal with unique predictors for an explicit reason: In two-equation systems, shared predictors do not aid in identification. This is because they do not provide a net increase in the information available to identify unknown parameters. The information brought to the model by a shared predictor in a two-equation system is just enough to identify the parameters associated with the shared predictor itself.

Figure 3.16 Case 8

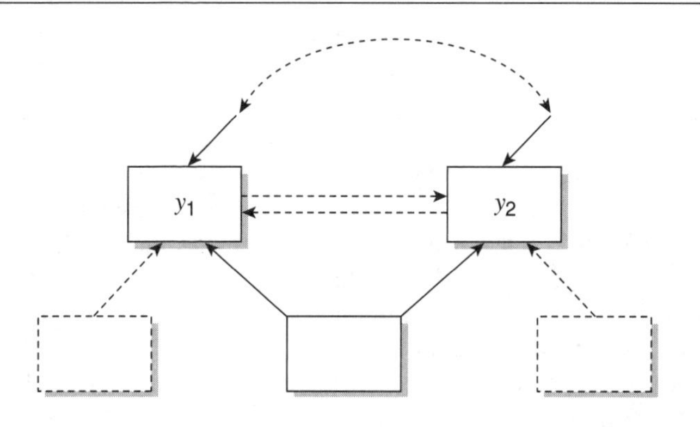

To classify a block with shared predictors, therefore, remove the shared predictor and reclassify the block as one of the cases 1 to 7. Of course, the shared predictor is not actually removed from the model in terms of specification or estimation, it is only removed in the identification step.

As an example, in the Ethington and Wolfle (1986) example displayed in Figure 3.5, the first block is as shown in Figure 3.17. This block would be reclassified as a Case 1 block once the two shared predictors are removed and is therefore identified. The Ethington and Wolfle (1986) model is therefore identified as we have demonstrated identification both within and across blocks.

Identifying the Duncan Example Using Rigdon's Eight Cases

We can also identify the Duncan et al. (1972) example using the eight special cases outlined by Rigdon. Figure 3.8 blocked the Duncan example into two blocks. The first block has two equations, y_1 and y_3, that are in a reciprocal relationship with correlated errors. Removing the shared predictors x_2 for x_3 leaves y_1 and y_3 with one unique predictor each: x_1 for y_1, and x_4 for y_3. This block can therefore be redrawn as Case 7 and is identified.

Figure 3.17 Ethington and Wolfe (1986), Block One

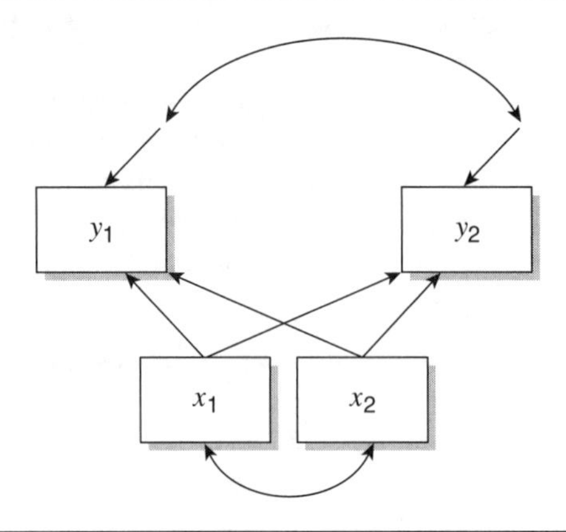

The second block is similarly identified through Case 7. In this block, y_2 and y_4 are in a reciprocal relationship with correlated errors. Removing the shared predictors x_2 and x_3 leaves y_2 and y_4 with two unique predictors each: x_1 and y_1 for y_2, and x_4 and y_3 for y_4. Note that this block would be identified even if x_1 and x_4 were not present to help identification. This is because y_1 and y_3 act as exogenous variables in the second block and therefore count as unique predictors. Readers are encouraged to look to Rigdon (1995) for additional examples.

Identifying the Voluntary Associations and Generalized Trust Example

We also illustrate identification using our exactly identified empirical example described in Chapter 2, shown in a path diagram in Figure 2.4. This model is nonrecursive due to the reciprocal relationship between y_1, voluntary association memberships, and y_2, generalized trust. In principle, we can approach the question of this model's identification using any of the strategies outlined in this chapter: covariance algebra, rules of identification, or block-recursive techniques. According to the block-recursive technique, we can classify the exactly identified model as "Case 7," which is identified (Rigdon, 1995). The variable x_2, education, influences both voluntary associations and generalized trust, and is a shared predictor, which does not aid in determining the identification status of the model. Each endogenous variable in the model has a unique predictor: x_1, children less than 6 years of age, influences voluntary association memberships but not trust, and x_3, burglary, is hypothesized to influence generalized trust, but not membership in voluntary associations. As shown in Figure 2.4, there is a correlation between the errors in the equations; according to Case 7, it is identified regardless of the presence of this correlation. The overidentified model is also identified by Case 7.

A Strategy for Identifying Nonrecursive Models

When faced with identifying a nonrecursive model, it may be difficult for a researcher to determine exactly where to start. There are a number of rules and techniques that might be helpful, but there are also considerations such as restrictions in the $\mathbf{\Psi}$ matrix or whether the model can be blocked.

A recommended sequence for identification of nonrecursive models therefore follows:

1. If $\mathbf{\Psi}$ is full, use the rank condition to identify the model.
2. If $\mathbf{\Psi}$ is not full and the model can be written as block recursive,
    a. if all blocks are one or two equations, then use the set of eight special cases to establish the identification status of each block and therefore the whole model; but
    b. if any blocks have three or more equations, see if each 3+ equation block passes the rank condition. If any block does not pass the rank condition and its $\mathbf{\Psi}$ matrix is not full, use algebraic identification to determine whether the block is identified.
3. If $\mathbf{\Psi}$ is not full and the model cannot be written as block recursive, see if it passes the rank condition. If it does not, use algebraic identification to determine whether the model is identified.

CHAPTER 4. ESTIMATION

Following identification, the next step in modeling is estimation. Estimators of nonrecursive models can be classified into two types: limited-information and full-information methods. Limited-information methods estimate one equation at a time, whereas full-information estimators take into account information from all equations in the system during estimation. Although many structural equation modeling (SEM) texts focus almost exclusively on a full-information strategy such as maximum likelihood (ML), we stress that a limited-information approach, such as two-stage least squares, can be advantageous in certain circumstances. Briefly, while full-information estimators produce asymptotically efficient parameter estimates, they are more sensitive to specification error, since error in one equation can be spread to parameter estimation in all equations of the system. Limited-information estimators are less likely to spread specification error among equations.[1]

We focus here on one limited-information estimator (two-stage least squares [2SLS]) and two full-information estimators (three-stage least squares [3SLS] and ML). Since complete treatments of the mechanics of various estimation techniques are discussed elsewhere (Bollen, 1989b; Greene, 2008; Wooldridge, 2002), this chapter focuses on (1) briefly introducing the three estimators, (2) demonstrating why ignoring possible endogeneity and estimating an ordinary least squares (OLS) regression equation will lead to biased estimates, and (3) providing the relative benefits and costs of either a limited-information or full-information estimation strategy. The empirical example of voluntary associations and trust illustrates differences in estimates between limited- and full-information estimation techniques.

Consequences of Endogeneity for OLS Estimation

A key question when estimating a nonrecursive simultaneous equation model is whether any assumptions of the classic linear regression model are violated. That is, should OLS be used as an estimator of nonrecursive models? For nonrecursive models, a key OLS assumption—that the regressors and the disturbance terms are uncorrelated (i.e., $\text{COV}(\mathbf{x}, \boldsymbol{\zeta}) = 0$)—is violated. Due to the nonrecursive specification, at least one explanatory variable will be correlated with the error. To illustrate this, consider the following two-equation system:

---

[1]It is possible for misspecification to spread in limited-information estimators if the misspecification in one equation of the model results in using inappropriate instrumental variables for another equation in the model.

$$y_1 = \beta_{12}y_2 + \gamma_{11}x_1 + \zeta_1 \quad (4.1)$$

$$y_2 = \beta_{21}y_1 + \gamma_{22}x_2 + \zeta_2 \quad (4.2)$$

A reciprocal relationship between y_1 and y_2 exists. Intuitively, ζ_1 is correlated with y_2 because ζ_1 affects y_1, which in turn affects y_2. And ζ_2 is similarly correlated with y_1. As further illustration, focus on the second equation. The question is whether COV(y_1, ζ_2) = 0; whether an explanatory variable in Equation 4.2, y_1, is correlated with the disturbance in that equation, ζ_2.

The reduced form of y_1 is

$$y_1 = \frac{1}{1 - \beta_{12}\beta_{21}}(\beta_{12}\gamma_{22}x_2 + \gamma_{11}x_1 + \beta_{12}\zeta_2 + \zeta_1)$$

or

$$y_1 = \Pi_{11}x_1 + \Pi_{12}x_2 + \zeta_1^* \quad (4.3)$$

with composite disturbance term

$$\zeta_1^* = \frac{\beta_{12}\zeta_2 + \zeta_1}{1 - \beta_{12}\beta_{21}}$$

Substituting the reduced form of y_1 into COV(y_1, ζ_2) produces

$$\text{COV}(y_1, \zeta_2) = \text{COV}(\Pi_{11}x_1 + \Pi_{12}x_2 + \zeta_1^*, \zeta_2)$$

x_1 and x_2 are exogenous and uncorrelated with the disturbance, leaving

$$\text{COV}(y_1, \zeta_2) = \text{COV}(\zeta_1^*, \zeta_2)$$

But the composite disturbance term from the reduced form is *not* uncorrelated with ζ_2:

$$\text{COV}(y_1, \zeta_2) = \text{COV}\left(\frac{\beta_{12}\zeta_2 + \zeta_1}{1 - \beta_{12}\beta_{21}}, \zeta_2\right) \neq 0$$

With a correlation between an explanatory variable and the disturbance, a key assumption in OLS estimation is violated. *OLS estimation on the structural equations of nonrecursive models will therefore yield biased and inconsistent estimates*. Use of OLS will inaccurately assess causal effects at any sample size.

It is worth noting that OLS estimation, while inappropriate for estimation of the structural equations, is appropriate for the reduced-form equations. Again, the reduced-form equation of y_1 is

$$y_1 = \frac{1}{1 - \beta_{12}\beta_{21}}(\beta_{12}\gamma_{22}x_2 + \gamma_{11}x_1 + \beta_{12}\zeta_2 + \zeta_1) \tag{4.4}$$

or

$$y_1 = \Pi_{11}x_1 + \Pi_{12}x_2 + \zeta_1^*$$

Since both x_1 and x_2 are exogenous variables,

$$\mathrm{COV}(x_1, \zeta_1^*) = \mathrm{COV}\left(x_1, \frac{\beta_{12}\zeta_2 + \zeta_1}{1 - \beta_{12}\beta_{21}}\right) = 0 \tag{4.5}$$

and

$$\mathrm{COV}(x_2, \zeta_1^*) = \mathrm{COV}\left(x_2, \frac{\beta_{12}\zeta_2 + \zeta_1}{1 - \beta_{12}\beta_{21}}\right) = 0 \tag{4.6}$$

Therefore, OLS estimation is appropriate for the reduced-form equations.

A Limited-Information Estimator Alternative: Two-Stage Least Squares

2SLS is an appropriate limited-information estimator for nonrecursive models. 2SLS belongs to a larger class of estimators known as instrumental variable (IV) estimators. Recall that an IV, z_1, must meet two criteria: $\mathrm{COV}(z_1, \zeta_1) = 0$ and $\mathrm{COV}(z_1, x_1) \neq 0$. That is, instruments must be uncorrelated with the disturbance and must be correlated with the problematic variable.[2] In nonrecursive models, the problematic variables will be the *y*s. IV estimators use instruments, which do not appear in the equation with the problematic variable, to "clean" the problematic variable from its association with the disturbance.

Exogenous variables serve as their own instruments in an equation, so a complete list of IVs for an equation includes all the exogenous variables in

---

[2]The second requirement, $\mathrm{COV}(z_1, x_1) \neq 0$, is actually an approximation that holds only if the problematic variable is the only explanatory variable in the equation. In most situations, there are other explanatory variables in the equation and the requirement is that z_1 must be partially correlated with the problematic variable once the other exogenous variables in the equation have been netted out (Wooldridge, 2002, p. 84).

the equation (included instruments) as well as the omitted variables identified as instruments for the problematic variable (excluded instruments). The excluded variables are sometimes referred to as "identifying instrumental variables."[3]

Where more than one IV is available for a problematic variable, 2SLS chooses the combination of IVs that is most highly correlated with the problematic variable and is thus efficient in the class of IV estimators. Also, unlike IV estimation in other contexts, in nonrecursive systems of equations, the excluded IVs typically come from other equations in the system. For example, in Equations 4.1 and 4.2 above, x_1 can serve as an instrument for y_1, and x_2 can serve as an instrument for y_2.

In theory, 2SLS implements OLS estimation in two stages (hence the name). In the first stage, each endogenous independent variable in an equation is regressed on all exogenous variables from all equations in the system. That is, it is regressed on all exogenous variables from its equation (the included instruments) as well as the exogenous variables from other equations (the excluded instruments). For example, in Equation 4.1, the first stage of 2SLS would regress y_2 on x_1 and x_2.

In most cases, the first-stage equation will be the reduced-form equation. The first-stage regression for the first structural equation of the example above is

$$y_2 = \Pi_{21}x_1 + \Pi_{22}x_2 + \zeta_2^* \tag{4.7}$$

We obtain the predicted values $(\hat{y}_2)$ from the reduced-form equation

$$\hat{y}_2 = \hat{\Pi}_{21}x_1 + \hat{\Pi}_{22}x_2 \tag{4.8}$$

where the predicted value, $\hat{y}_2$, can now be considered the combined instrument for y_2.[4] The predicted value, $\hat{y}_2$, can be interpreted as the part of y_2 that is uncorrelated with the disturbance. Thus, the predicted values, by being computed from only exogenous variables, are "cleaned" of their association with the error term. The key insight is that the first stage uses

[3]Confusion arises in terminology as some texts refer to the excluded instruments as the "instrumental variables" without acknowledging that the complete set of instrumental variables includes not only the instruments that are excluded, or omitted, from the equation but also the full set of exogenous variables in the equation.

[4]Again, confusion in terminology can arise. Some texts refer to $\hat{y}_2$ as the "instrument" for y_2. While technically correct, we choose here to call it the combined instrument to distinguish it from the excluded and included instrumental variables in the structural equations.

OLS estimation on the *reduced-form* equations. As discussed above, because the disturbance term is uncorrelated with the exogenous variables, OLS estimation can be used on the reduced-form equations.

In the second stage of 2SLS, we apply least squares to the structural equations using the fitted values of the endogenous variables generated in the first stage. In our example, we substitute $\hat{y}_2$ for the original y_2 in the structural equation:

$$y_1 = \beta_{12}\hat{y}_2 + \gamma_{11}x_1 + \zeta_1 \tag{4.9}$$

This equation no longer contains a problematic variable (i.e., no explanatory variables are correlated with the disturbance) and can be estimated with OLS. Note that the disturbance is more complicated here as it contains the leftover part of y_2.

Actually performing 2SLS by hand requires that the standard errors be corrected to account for the uncertainty in this predicted value. As this correction is carried out by various statistical programs, it is best to estimate 2SLS using a software package.

We can represent 2SLS estimation more generally in nonrecursive simultaneous equation models. Consider the following single equation from a set of structural equations:

$$y_i = \mathbf{Y}_i\boldsymbol{\beta}_i + \mathbf{X}_i\boldsymbol{\gamma}_i + \boldsymbol{\zeta}_i$$

where the i subscript indicates a single equation of p total possible equations. $\mathbf{Y}_i \neq \mathbf{Y}$ and $\mathbf{X}_i \neq \mathbf{X}$, meaning that not all variables in the complete system appear in the equation.

Define

$$\mathbf{Z} = [\mathbf{Y}_i\ \mathbf{X}_i]$$

and

$$\boldsymbol{\delta}_i = \begin{bmatrix} \boldsymbol{\beta}_\text{i} \\ \boldsymbol{\gamma}_\text{i} \end{bmatrix}$$

so that all coefficients appear in a single vector, leading to

$$y_i = \mathbf{Z}_i\boldsymbol{\delta}_i + \zeta_i. \tag{4.10}$$

Remember that $\mathbf{Y}_i$ is correlated with the error term.

In 2SLS, instead of $\mathbf{Y}_i$, $\hat{\mathbf{Y}}_i$ is used, where the instruments are all exogenous variables ($\mathbf{X}$) in the system of equations.[5]

$$\hat{\mathbf{Y}}_i = \mathbf{X}[(\mathbf{X}'\mathbf{X})^{-1}\mathbf{X}'\mathbf{Y}_i] \tag{4.11}$$

From this, we generate the following equation:

$$y_i = \hat{\mathbf{Z}}_i\boldsymbol{\delta}_i + \zeta_i^*$$

where $\hat{\mathbf{Z}}_i = [\hat{\mathbf{Y}}_i \; \mathbf{X}_i]$ and ζ_i^* is now asymptotically uncorrelated with $\hat{\mathbf{Z}}_i$.

The IV estimator in 2SLS is thus

$$\begin{aligned} \hat{\boldsymbol{\delta}}_{i,\text{2SLS}} = \begin{bmatrix} \hat{\boldsymbol{\beta}}_{i,\text{2SLS}} \\ \hat{\boldsymbol{\gamma}}_{i,\text{2SLS}} \end{bmatrix} &= \left[(\hat{\mathbf{Z}}'_i\hat{\mathbf{Z}}_i)^{-1}\hat{\mathbf{Z}}'_i y_i\right] \\ &= \begin{bmatrix} \hat{\mathbf{Y}}'_i\hat{\mathbf{Y}}_i & \hat{\mathbf{Y}}'_i\mathbf{X}_i \\ \mathbf{X}'_i\hat{\mathbf{Y}}_i & \mathbf{X}'_i\mathbf{X}_i \end{bmatrix}^{-1} \begin{bmatrix} \hat{\mathbf{Y}}'_i y_i \\ \mathbf{X}'_i y_i \end{bmatrix} \end{aligned} \tag{4.12}$$

Ordinary Least Squares Versus Two-Stage Least Squares

Like all IV estimators, 2SLS can have finite sample bias. But 2SLS is asymptotically unbiased, consistent, and asymptotically efficient among all limited-information IV estimators.[6]

When estimating a nonrecursive model, OLS is biased in small samples and typically has the largest bias of all commonly employed estimators. More important, OLS estimation is inconsistent. OLS also provides standard errors that are too small, which has the consequence of rejecting the null hypothesis too frequently (Kmenta, 1997, p. 714). The 2SLS estimator also has finite sample bias, though its bias tends to be smaller than OLS. OLS and 2SLS tend to be biased in the same direction. 2SLS does have a larger variance than OLS. In fact, based on a mean square error criterion (a combination of bias and variance), it might sometimes be preferable to

---

[5]Some are included in the equation and some excluded from the equation. Note that in some models $\mathbf{X}$ would include variables endogenous in other equations. This can occur if there are some zero correlations in the model.

[6]Even in large samples, bias can occur if instruments are weak (see Chapter 5).

use OLS rather than 2SLS, although this would be a potential strategy only in very small samples.

2SLS standard errors tend to be larger than OLS, with the size of the standard errors depending on the quality of the IVs. If the R^2 from the first-stage regression is small, it causes a larger asymptotic variance for the coefficient estimates, which can lead to insignificance in 2SLS estimates. We will discuss the issues of weak IVs, and how to test for them, in Chapter 5.

Given the challenges associated with identifying suitable IVs, and the need to test their validity, some researchers may be tempted to simply ignore possible endogeneity in their model. As highlighted above, OLS is inconsistent and will not produce correct parameter estimates at any sample size. In Table 4.1, we compare the OLS and 2SLS estimates for the exactly identified empirical example that we described in Chapter 2. Recall that

Table 4.1 Comparison of OLS and 2SLS, Exactly Identified Example

	(1)	*(2)*
	2SLS	*OLS*
Association memberships outcome		
Interpersonal trust	0.027	0.191
	(0.309)	(0.031)
Years of education	0.184	0.174
	(0.021)	(0.008)
Presence of children less than 6 years of age	−0.189	−0.167
	(0.060)	(0.044)
Intercept	−0.405	−0.266
	(0.284)	(0.111)
R^2	0.108	0.113
Interpersonal trust outcome		
Association memberships	0.679	0.046
	(0.179)	(0.007)
Years of education	−0.064	0.052
	(0.034)	(0.004)
Experienced a burglary in the past year	−0.300	−0.310
	(0.075)	(0.045)
Intercept	−0.531	−0.816
	(0.118)	(0.052)
R^2	—	0.071

NOTE: N = 4,598; OLS = ordinary least squares; 2SLS = two-stage least squares.

voluntary associations and trust are modeled in a reciprocal relationship. This model is exactly identified with one IV for each of our endogenous variables and one measure (education) predicting both of these outcomes.

The results are considerably different using OLS and 2SLS. In the OLS estimates, association membership has a positive relationship with interpersonal trust, *and* interpersonal trust significantly increases one's memberships. In 2SLS, only the membership to interpersonal trust relationship is significant. In OLS, the positive relationship between these two measures does not get partialled into that attributable to each of the variables. Instead, it is attributed simply to whichever of these measures we have specified as exogenous in the equation. In other words, OLS incorrectly assigns a portion of the variation in the outcome variable generated by the disturbance to the endogenous variable with which the disturbance is correlated. These results also demonstrate that ignoring endogeneity can affect the coefficients of other variables in the model. Note also that the standard errors in these OLS equations are smaller than those in the 2SLS equations, which reflects the increased variance that comes from a 2SLS approach. Finally, it is possible to obtain a negative R^2 value using 2SLS. See Chapter 5 for a discussion of this phenomenon. We follow the increasingly common practice of not reporting a negative R^2—as discussed in Chapter 5, the interpretation of R^2 is not straightforward in nonrecursive models.

Full-Information Estimation: Three-Stage Least Squares

We next discuss two full-information estimators: 3SLS and full-information maximum likelihood (ML). 3SLS is a full-information estimation technique that uses information from other equations in the system in the estimation of structural parameters (Zellner & Theil, 1962). The 3SLS estimator is not as commonly used as either 2SLS or ML, but it builds on the 2SLS estimator and is therefore useful to briefly describe.

The logic of 3SLS is that efficiency gains are possible by applying generalized least squares (GLS) to the 2SLS equations. Begin by stacking the second-stage equations from 2SLS:

$$
\begin{aligned}
y_1 &= \hat{\mathbf{Z}}_1\boldsymbol{\delta}_1 + \zeta_1^* \\
y_2 &= \hat{\mathbf{Z}}_2\boldsymbol{\delta}_2 + \zeta_2^* \\
&\vdots \\
y_i &= \hat{\mathbf{Z}}_i\boldsymbol{\delta}_i + \zeta_i^*
\end{aligned}
$$

where $\hat{\mathbf{Z}}_i = \begin{bmatrix} \hat{\mathbf{Y}}_i & \mathbf{X}_i \end{bmatrix}$.

Alternatively,

$$\mathbf{y} = \hat{\mathbf{Z}}\boldsymbol{\delta} + \boldsymbol{\zeta}^{*} \tag{4.13}$$

We can apply GLS to this equation, resulting in

$$\hat{\boldsymbol{\delta}}_{\text{GLS}} = \left[\hat{\mathbf{Z}}'\left(\boldsymbol{\Psi}^{-1} \otimes \mathbf{I}\right)\hat{\mathbf{Z}}\right]^{-1}\left[\hat{\mathbf{Z}}'\left(\boldsymbol{\Psi}^{-1} \otimes \mathbf{I}\right)\mathbf{y}\right] \tag{4.14}$$

where the variance-covariance matrix of $\boldsymbol{\zeta}$ is $\boldsymbol{\Psi} \otimes \mathbf{I}$.

$\boldsymbol{\Psi}$ must be estimated. Zellner and Theil (1962) suggest using the residuals from the 2SLS estimates.

$$\psi_{ij} = \frac{\left(y_i - \hat{\mathbf{Z}}_i\hat{\boldsymbol{\delta}}_i\right)'\left(y_j - \hat{\mathbf{Z}}_j\hat{\boldsymbol{\delta}}_j\right)}{N} \tag{4.15}$$

Substituting $\hat{\boldsymbol{\Psi}}$ for $\boldsymbol{\Psi}$ in Equation 4.14, we obtain the 3SLS estimator:

$$\hat{\boldsymbol{\delta}}_{\text{3SLS}} = \left[\hat{\mathbf{Z}}'\left(\hat{\boldsymbol{\Psi}}^{-1} \otimes \mathbf{I}\right)\hat{\mathbf{Z}}\right]^{-1}\left[\hat{\mathbf{Z}}'\left(\hat{\boldsymbol{\Psi}}^{-1} \otimes \mathbf{I}\right)\mathbf{y}\right] \tag{4.16}$$

In short, (1) residuals from the 2SLS second-stage (structural) equations are calculated, (2) these residuals are used to estimate the variance/covariance matrix of the equation disturbances, and (3) that matrix is then used as the "sandwich" matrix in a GLS estimation on the stacked set of equations.

It follows that there are two situations where 2SLS and 3SLS will yield identical results. The first situation is when the covariances among the equation disturbances are all zero. The second is when all the equations are exactly identified (Zellner & Theil, 1962). In both instances, there is no net gain in information for generating estimates between the two techniques. Essentially, exactly identified equations add no new information to the system. Thus, with two equations in a model, one exactly identified and one overidentified, the overidentified equation's estimates will be the same in 3SLS and 2SLS. This is because the exactly identified equation is not adding any new information to the estimate of the overidentified equation. The exactly identified equation will have estimates that differ from 2SLS, however, because the restrictions from the overidentified equation will be used in its estimation during 3SLS. If, however, there is more than one overidentified equation in a model, then all equations' estimates would differ between 3SLS and 2SLS.

Full-Information Maximum Likelihood

The most widely used full-information estimator of simultaneous equation models is the ML estimator. The ML estimator treats all equations and all parameters jointly in attempting to minimize the difference between the elements of the model-implied covariance matrix and those of the population covariance matrix: $\mathbf{\Sigma} = \mathbf{\Sigma}(\boldsymbol{\theta})$. Using existing data, the model's parameters are statistically estimated using the sample realizations of these two matrices, **S** and $\mathbf{\Sigma}(\hat{\boldsymbol{\theta}})$. The parameters of a model include (1) elements of $\mathbf{\Phi}$ (the covariance matrix of exogenous variables), (2) elements in $\mathbf{\Psi}$ (the covariance matrix of disturbance terms), and (3) coefficients contained in $\mathbf{\Gamma}$ and **B**. All these matrices appear in $\mathbf{\Sigma}(\boldsymbol{\theta})$, and the matrices containing their estimates, $\hat{\mathbf{\Phi}}, \hat{\mathbf{\Psi}}, \hat{\mathbf{B}},$ and $\hat{\mathbf{\Gamma}}$, appear in $\mathbf{\Sigma}(\hat{\boldsymbol{\theta}})$.

Based on an SEM with latent variables approach, the ML fitting function can be written as

$$F_{\mathrm{ML}} = \log|\mathbf{\Sigma}(\boldsymbol{\theta})| + tr\left(\mathbf{S}\mathbf{\Sigma}^{-1}(\boldsymbol{\theta})\right) - \log|\mathbf{S}| - (p + q) \qquad (4.17)$$

Derivations of this estimator are shown elsewhere (Bollen, 1989b, pp. 131–135). To understand this fitting function, consider the situation of perfect fit, where $\mathbf{\Sigma}(\hat{\boldsymbol{\theta}}) = \mathbf{S}$. This situation could occur in the rare instance when a model does fit the data perfectly. More often, it will occur when your model is just-identified. When $\mathbf{\Sigma}(\hat{\boldsymbol{\theta}}) = \mathbf{S}$, the first and third terms of Equation 4.17 are equivalent and of opposite signs. Also, the second and fourth terms are equivalent and of opposite signs $(\mathbf{S}\mathbf{\Sigma}^{-1}(\hat{\boldsymbol{\theta}}) = \mathbf{I}_{p+q})$. Altogether, a perfect correspondence between these covariance matrices results in a value of zero for the ML fitting function.

In most overidentified situations, the ML fitting function will not result in a value of zero, and the farther apart **S** is from $\mathbf{\Sigma}(\hat{\boldsymbol{\theta}})$, the larger the value of F_{ML}.

Understanding Iterated Estimation

ML estimation uses a fitting function to iterate toward estimates of the unknown population parameters in $\mathbf{\Phi}$, $\mathbf{\Psi}$, **B**, and $\mathbf{\Gamma}$. In this section, we provide an intuitive sense of how iterated estimation works.

Recall the example of a three-variable model in Chapter 2 with the population observed covariance matrix of

$$\mathbf{\Sigma} = \begin{bmatrix} \mathrm{VAR}(y_1) & & \\ \mathrm{COV}(y_2, y_1) & \mathrm{VAR}(y_2) & \\ \mathrm{COV}(x_1, y_1) & \mathrm{COV}(x_1, y_2) & \mathrm{VAR}(x_1) \end{bmatrix}$$

and the model-implied covariance matrix of

$$\boldsymbol{\Sigma}(\boldsymbol{\theta}) = \begin{bmatrix} \gamma_{11}^2\phi_{11} + \psi_{11} & & \\ (\gamma_{11}{}^2\beta_{21} + \gamma_{11}\gamma_{21})\phi_{11} + \beta_{21}\psi_{11} & (\beta_{21}^2\gamma_{11}^2\phi_{11} + 2\beta_{21}\gamma_{11}\gamma_{21} + \gamma_{21}^2)\phi_{11} + \beta_{21}^2\psi_{11} + \psi_{22} & \\ \gamma_{11}\phi_{11} & \beta_{21}\gamma_{11}\phi_{11} + \gamma_{21}\phi_{11} & \phi_{11} \end{bmatrix}$$

In estimation, the goal is to minimize the differences between these covariance matrices, $\boldsymbol{\Sigma} = \boldsymbol{\Sigma}(\boldsymbol{\theta})$, although we are working with the sample counterparts, **S** and $\boldsymbol{\Sigma}(\hat{\boldsymbol{\theta}})$. In the example above, suppose **S** is

$$\mathbf{S} = \begin{bmatrix} 14 & & \\ 11 & 25 & \\ 6 & 9 & 2 \end{bmatrix}$$

Suppose we begin with guesses for the parameters of $\boldsymbol{\Phi}$, $\boldsymbol{\Psi}$, **B**, and $\boldsymbol{\Gamma}$, that lead to a $\boldsymbol{\Sigma}(\hat{\boldsymbol{\theta}})$ of

$$\boldsymbol{\Sigma}(\hat{\boldsymbol{\theta}}) = \begin{bmatrix} 13 & & \\ 11 & 24 & \\ 6 & 8 & 2 \end{bmatrix}$$

The difference between values in **S** and $\boldsymbol{\Sigma}(\hat{\boldsymbol{\theta}})$ is the residual matrix, which shows how close the two covariance matrices are. We could change our estimates of $\boldsymbol{\Phi}$, $\boldsymbol{\Psi}$, **B**, and $\boldsymbol{\Gamma}$ slightly to change $\boldsymbol{\Sigma}(\hat{\boldsymbol{\theta}})$ and possibly reduce the values in the residual matrix.

The ML estimator, based on the constraints specified in the model-implied covariance matrix, iterates a number of times in this fashion toward an optimal solution. A particular threshold for the change in the log likelihood is chosen as a stopping criterion, and estimation concludes once the improvement in the log likelihood is less than this criterion for a particular iteration. See Bollen (1989b, pp. 136–144) for a clear description of numerical solutions to minimize fitting functions.

One benefit of obtaining the fitting function during estimation is that it provides a measure of the overall goodness of fit of the model in the case of overidentification. This and other measures of fit will be discussed in Chapter 5.

Full-Information Versus Limited-Information Estimation

A key difference between full-information and limited-information estimators is a trade-off between bias and variance. Full-information methods take advantage of all the information from all equations specified in the simultaneous equation system, whereas single-equation estimators ignore the correlation between equation disturbances and therefore the restrictions that appear in other equations in the system. Full-information estimators, therefore, have a large-sample efficiency advantage.[7] The degree of this efficiency gain will depend on the particular model, sample size, and kurtosis in the observed variables. The efficiency gain of full relative to limited-information estimation methods is also smaller when the correlations between disturbances are small.[8] Some simulation results have found only modest improvements for ML versus 2SLS among reasonable model conditions (Bollen, Kirby, Curran, Paxton, & Chen, 2007).

Although full-information techniques can be more efficient than limited-information techniques, they are more vulnerable to model misspecification (Bollen et al., 2007; Cragg, 1968). Although full-information techniques are consistent when the model is correctly specified, any misspecification in the model can bias parameter estimates throughout the model (Curran, 1994; Kaplan, 1988). In contrast, given that 2SLS estimates each equation separately, misspecification in one equation will not affect the parameter estimates in other equations (as long as it does not affect the excluded variables/instruments) (Bollen, 1996). This allows the researcher to isolate potential model misspecification (Kirby & Bollen, 2009). Overall, there is some simulation evidence that the magnitude of the costs of bias due to model misspecification for full-information techniques is much greater than the size of the gains due to greater efficiency (Bollen et al., 2007). This is an area of ongoing research.

The estimation techniques discussed here (2SLS, 3SLS, and ML) are all consistent. Thus, as the sample size increases, their estimates will all approach the population parameters. It is also the case that all the estimation techniques discussed here will suffer from small sample bias. However, the simulation literature shows no clear way to rank limited- and full-information estimators in the extent of their small sample bias.

The researcher is therefore cautioned to carefully select an appropriate estimation technique, recognizing the trade-offs between limited- and full-information techniques. Limited-information estimators such as 2SLS are *not* outdated methods that can be safely ignored by researchers using structural

[7]The ML estimator is consistent, asymptotically unbiased, and asymptotically efficient when the model is correctly specified and the observed variables do not have excessive kurtosis. For normally distributed disturbances, 3SLS and ML have the same asymptotic distribution. In fact, it is possible to produce numerically equivalent results for 3SLS and ML by iterating 3SLS.

[8]The situation is parallel to the appeal of seemingly unrelated regression models, or SUR models, where ignoring the correlated disturbances by using OLS limits efficiency. Also similar to the OLS/SUR comparison, full-information and limited-information estimators of nonrecursive models will be equivalent for exactly identified models.

equation software packages. Indeed, unless the researcher is very confident in the specification of all equations in the model, a case can be made for the limited-information estimator in an effort to isolate any misspecifications. Moreover, there is a wealth of information contained in the reduced-form equations of nonrecursive models that is often opaque to researchers adopting a full-information approach, as we describe in the next chapter.

Empirical Example: Exactly Identified Case

We illustrate these various estimators with the empirical examples introduced in Chapter 2. Recall that voluntary associations and trust are modeled in a reciprocal relationship. The first model is exactly identified with one IV for each of our endogenous variables, and one measure (education) predicting both of these outcomes. Table 4.2 presents results for three estimators: (1) the 2SLS estimator, (2) the 3SLS estimator, and (3) the ML estimator. Because the model is exactly identified, there is only one possible solution; thus, all these estimators return the same results.

Table 4.2 Results for Exactly Identified Model, Using 2SLS, 3SLS, and ML

Association memberships outcome	
Interpersonal trust	0.027 (0.309)
Years of education	0.184 (0.021)
Presence of children less than 6 years of age	–0.189 (0.060)
Intercept	–0.405 (0.284)
R^2	0.108
Interpersonal trust outcome	
Association memberships	0.679 (0.179)
Years of education	–0.064 (0.034)
Experienced a burglary in the past year	–0.300 (0.075)
Intercept	–0.531 (0.118)
R^2	—

NOTE: N = 4,598; 2SLS = two-stage least squares; 3SLS = three-stage least squares; ML = maximum likelihood.

For all three estimators, we obtain the same coefficient estimates and the same standard errors. This emphasizes the point made earlier that efficiency gains, if there are any, will only be achieved with overidentified models.

Of substantive interest in the model was the possible reciprocal relationship between voluntary association memberships and interpersonal trust. In this model, there is no evidence that levels of interpersonal trust affect memberships, as the estimated coefficient is smaller than the standard error, suggesting a nonsignificant finding. On the other hand, there is evidence that individuals reporting greater levels of memberships report more interpersonal trust: Each additional organization type that a person reports belonging to is associated with a 0.68 increase in interpersonal trust, when controlling for a person's level of education and having experienced a burglary in the past year.

Other variables in the model show the expected effects. For instance, as hypothesized, the presence of young children has a significant negative effect on voluntary association memberships and experiencing a burglary in the past year significantly reduces interpersonal trust. This is one diagnostic for our IVs. We will discuss others in Chapter 5. Finally, persons with higher levels of education report higher levels of memberships, consistent with our expectations. Higher levels of education, however, are only marginally significantly related to interpersonal trust.

Syntax for Stata and SAS for Nonrecursive Models

There are slight differences in Stata code for 2SLS estimation between versions starting with 10, and those prior to 10. We show the code beginning with Stata 10:

```
*voluntary associations equation
ivregress assoc (intprtrst = burglary) educ babies, first
*interpersonal trust equation
ivregress intprtrst (assoc = babies) educ burglary, first
```

In this code, *assoc* is the voluntary associations variable, *intprtrst* is interpersonal trust, *burglary* is an indicator for experiencing a burglary in the past year, *babies* indicates whether there are children less than 6 years of age in the household, and *educ* is years of education for the respondent.

For versions prior to Stata 10, this code can be modified by replacing **ivregress** with **ivreg2**. Although **ivreg** is the built-in command for 2SLS estimation pre-Stata 10, the **ivreg2** command has several nice additional features for assessing the quality of the IVs and can be downloaded and installed into Stata by typing "**install ivreg2**."

For the voluntary associations equation, we denote the outcome variable first (*assoc*). We then denote the endogeneous variables in the equation in parentheses (*intprtrst*), and after the equal sign, the excluded IVs for this equation are listed: These are variables that do not appear in this equation (*burglary*) but are hypothesized to predict this endogenous variable. We then list the rest of the exogenous variables in the equation predicting voluntary associations (*educ* and *babies*).

The logic is the same for the interpersonal trust equation. First, we list this outcome variable (*intprtrst*), then in parentheses before the equal sign, we list the endogenous variables that are predictors in this equation (*assoc*), and after the equal sign, we list the excluded IVs (*babies*). The last part of the line lists the rest of the exogenous variables in the equation (*educ* and *burglary*).

Any commands listed after the comma are options. In this case, we are asking Stata to provide the results from the first-stage equations. Note that Stata can provide heteroscedastic-robust standard errors with the command **robust**.

SAS code for 2SLS estimation is as follows:

```
proc syslin data=a1 2sls first;
endogenous assoc intprtrst;
instruments burglary educ babies;
assoc: model assoc = intprtrst educ babies/overid;
run;

proc syslin data=a1 2sls first;
endogenous assoc intprtrst;
instruments burglary educ babies;
intprtrst: model intprtrst = assoc educ burglary/overid;
run;
```

These commands use the **syslin** procedure (system of linear equations). The **2sls** command on the first line tells SAS to use 2SLS estimation, and the **first** option prints out the results of the first stage equation(s). The **endogenous** command lists all variables that are endogenous in the system. The **instruments** command lists *all* exogenous variables in the system. We name the model using the notation: "assoc:". After the **model** command, we list the outcome variable (assoc) before the equal sign, and then all predictors in the equation after the equal sign, including the endogenous variable first in the list. After the "/" symbol we can list various options; here we are asking for various overidentification tests (**overid**), which will be described in Chapter 5. The interpersonal trust equation is specified similarly, with the difference that the model statement lists the variables included in this equation.

Stata code for 3SLS estimation is

```
reg3 (assoc intprtrst educ babies) (intprtrst assoc educ burglary)
```

The **reg3** command tells Stata to use 3SLS estimation. The equations in the system are specified within the sets of parentheses; in each case, the outcome variable is listed first, and all other variables in the equation (endogenous or exogenous) are listed after that.

SAS code for 3SLS estimation is as follows:

```
proc syslin data=a1 3sls first;
endogenous assoc intprtrst;
instruments burglary educ babies;
assoc: model assoc = intprtrst educ babies/overid;
intprtrst: model intprtrst = assoc educ burglary/overid;
run;
```

We again use the **syslin** procedure, but the **3sls** command on the first line tells SAS to use 3SLS estimation. The **endogenous** command again lists all endogenous variables in the system, whereas the **instruments** command lists all exogenous variables in the system. The **model** statements specify the actual structural equations.

SAS code for ML estimation using proc syslin is as follows:

```
proc syslin data=a1 fiml first;
endogenous assoc intprtrst;
instruments burglary educ babies;
assoc: model assoc = intprtrst educ babies/overid;
intprtrst: model intprtrst = assoc educ burglary/overid;
run;
```

Within the **syslin** procedure, the **fiml** command on the first line tells SAS to use ML estimation. The **endogenous** command again lists all endogenous variables in the system, whereas the **instruments** command lists all exogenous variables in the system. The **model** statements specify the actual structural equations.

Alternatively, the researcher can use the **proc calis** command in SAS, which also uses ML estimation to obtain identical results to proc syslin. One benefit of using the calis procedure is that it provides overall goodness-of-fit statistics for the model, as will be discussed in Chapter 5. Here is the code to estimate this particular model:

```
proc calis data = a1 method=ml platcov ucov aug;
var assoc intprtrst educ babies burglary;
lineqs
assoc = a11 intercept + b12 intprtrst + g11 educ + g12 babies + d1,
intprtrst = a21 intercept + b21 assoc + g21 educ + g23 burglary + d2;
std
d1-d2 = ph1-ph2;
cov
d1-d2 = ph12;
run;
```

We use the **calis** procedure to estimate this system of equations with ML. The **method=ML** tells SAS to use the ML estimator. The **aug** subcommand allows us to specify intercepts in the equations. The **var** line lists all variables in the entire model. The **lineqs** section lists the equations of the system. For the voluntary associations equation, we first list the outcome (*assoc*), and then after the equal sign, we list the variables in the equation along with the intercept. We must also define a unique parameter for each variable. Thus, "a11" represents the parameter for the intercept in this equation, "b12" the parameter for the effect of *intprtrst* on *assoc*, "g11" the parameter for the effect of *educ* on *assoc*, "g12" the parameter for *babies* on *assoc*, and "d1" the disturbance. Note that the last equation in this section ends with a semicolon (;), and all other equations end with commas. Note also that in the interpersonal trust equation, the parameters are given unique names to distinguish them from the parameters in the voluntary associations equation.

The **std** section allows us to define the variances of the disturbances. Here, we define the variances of the variables, d1 and d2, to be "ph1" and "ph2." The **cov** section allows us to define covariances. We define the covariance between the two disturbances to be "ph12." In other words, we explicitly include the correlation between the disturbance terms in the system of simultaneous models as specified in Figure 2.4.

Empirical Example: Overidentified Case

Although the estimator used makes no difference for the results in an exactly identified model, this is not the case for an overidentified model. To reiterate, for this model, we modify the previous model by adding a second excluded IV to each equation. Hours of TV viewing is modeled as influencing association participation, and having experienced the divorce of one's parents is modeled as influencing trust. The results of this overidentified model are presented in Table 4.3. Results show it is now possible to observe different estimates for the different estimators.

First, for our key research question regarding the reciprocal relationship between voluntary associations and interpersonal trust, our substantive conclusion is the same as in the exactly identified model. Regardless of the estimator, there is no evidence that higher levels of interpersonal trust lead to more memberships, as this estimate is always less than the standard error. On the other hand, there is consistent evidence that memberships increase reported interpersonal trust: The effect of participation in each additional organization type on interpersonal trust ranges from 0.472 in the 3SLS estimator to 0.474 in the 2SLS estimator to 0.490 for the ML estimator.

Table 4.3 Results for Overidentified Model, Using 2SLS, 3SLS, and ML

	2SLS Estimation	*3SLS Estimation*	*ML Estimation*
Association memberships outcome			
Interpersonal trust	0.076	0.067	0.070
	(0.275)	(0.275)	(0.275)
Years of education	0.168	0.170	0.170
	(0.018)	(0.018)	(0.017)
Hours of TV viewing	–0.066	–0.059	–0.058
	(0.014)	(0.013)	(0.013)
Presence of children less than 6 years of age	–0.178	–0.214	–0.210
	(0.056)	(0.050)	(0.049)
Intercept	–0.002	–0.043	–0.049
	(0.230)	(0.228)	(0.223)
R^2	0.117	0.116	0.116
Interpersonal trust outcome			
Association memberships	0.474	0.472	0.490
	(0.090)	(0.090)	(0.092)
Years of education	–0.025	–0.024	–0.028
	(0.017)	(0.017)	(0.018)
Experienced a burglary in the past year	–0.298	–0.283	–0.282
	(0.060)	(0.056)	(0.057)
Parents divorce at age 16	–0.072	–0.084	–0.084
	(0.032)	(0.025)	(0.025)
Intercept	–0.605	–0.604	–0.596
	(0.080)	(0.080)	(0.082)
R^2	—	—	—

NOTE: N = 4,598; 2SLS = two-stage least squares; 3SLS = three-stage least squares; ML = maximum likelihood.

For other measures in the model, there are differences based on the estimator. For example, in the equation with interpersonal trust as the outcome, the effect of experiencing a burglary in the past year has a slightly stronger negative effect with the 2SLS estimator compared with the other two estimators. In contrast, the effect of having divorced parents at age 16 has a weaker effect with the 2SLS estimator. Likewise, there are some differences in the civic participation equation, as the effect of television viewing is stronger with the 2SLS estimator compared with the effect of the presence of young children, which is weaker with the 2SLS estimator. These are all extremely modest differences.

Stata code for 2SLS estimation is as follows:

```
*voluntary associations equation
ivregress assoc (intprtrst = burglary pardiv16) educ tvhours babies, first
*interpersonal trust equation
ivregress intprtrst (assoc = babies tvhours) educ burglary pardiv16, first
```

SAS code for 2SLS estimation is as follows:

```
proc syslin data=a1 2sls first;
endogenous assoc intprtrst;
instruments burglary educ babies tvhours pardiv16;
assoc: model assoc = intprtrst educ babies tvhours/overid;
run;
proc syslin data=a1 2sls first;
endogenous assoc intprtrst;
instruments burglary educ babies tvhours pardiv16;
intprtrst: model intprtrst = assoc educ burglary pardiv16/overid;
run;
```

Stata code for 3SLS estimation is as follows:

```
reg3 (assoc intprtrst educ tvhours babies) (intprtrst assoc educ
burglary pardiv16)
```

SAS code for 3SLS estimation is as follows:

```
proc syslin data=a1 3sls first;
endogenous assoc intprtrst;
instruments burglary educ babies tvhours pardiv16;
assoc: model assoc = intprtrst educ babies tvhours/overid;
intprtrst: model intprtrst = assoc educ burglary pardiv16/overid;
run;
```

SAS code for ML estimation is as follows:

```
proc syslin data=a1 fiml first;
endogenous assoc intprtrst;
instruments burglary educ babies tvhours pardiv16;
assoc: model assoc = intprtrst educ babies tvhours/overid;
intprtrst: model intprtrst = assoc educ burglary pardiv16/overid;
run;
```

Using the **calis** procedure to estimate this system of equations with ML requires the following code:

```
proc calis data = a1 method=ml platcov ucov aug;
var assoc intprtrst educ babies burglary tvhours pardiv16;
lineqs
assoc = a11 intercept + g12 intprtrst + g13 educ + g15 tvhours
+ g14 babies + d1,
intprtrst = a21 intercept + g21 assoc + g23 educ + g24 burglary
+ g25 pardiv16 + d2;
std
d1-d2 = ph1-ph2;
cov
d1-d2 = ph12;
run;
```

CHAPTER 5. ASSESSMENT

In this chapter, we discuss how to assess the model after estimation. We focus on (1) assessing the component fit of each individual equation, (2) assessing overall fit of the system of equations in the tradition of structural equation modeling (SEM) with latent variables, and (3) assessing the quality of the instrumental variables (IVs). The fact that SEM software packages using maximum likelihood estimation rarely provide the results of tests of the IVs may obfuscate the importance of assessing their quality. But these tests are critical no matter what estimator is used—bad or weak IVs undermine the properties of any estimator. The tests we describe are not difficult to implement and can be conducted in most standard statistical software packages (e.g., SAS, Stata).

Assessing Individual Equations

Each equation in the multiple equation system should be assessed using the same diagnostics used when estimating a single equation with ordinary least squares (OLS; Belsley, Kuh, & Welsch, 2004; Fox, 1991; Long, 1988). Given that multicollinearity, heteroscedasticity, outliers, and so on are extensively covered in texts focused on OLS models, we will not cover them here.[1] However, researchers should recognize that the overall assessment of the model rests first on the quality of each of the individual equations. For example, researchers should use their substantive knowledge to determine whether the coefficient estimates have appropriate signs, are significant, and so on. Researchers will also want to assess whether the variance explained for each individual equation (the R^2) is in line with acceptable levels for their substantive area.

However, the interpretation of R^2 is not straightforward in nonrecursive models. To understand the complication, consider one definition of R^2, the squared multiple correlation

$$R_i^2 = 1 - \mathrm{VAR}(\zeta_i)/\mathrm{VAR}(y_i) \tag{5.1}$$

where ζ_i is the disturbance term in a structural equation, and y_i is the outcome variable in that same equation. The usual interpretation is the

[1]Some of these diagnostic techniques have analogues in structural equation modeling (Bollen, 1996; Bollen & Kmenta, 1986).

amount of variance in y explained by or accounted for by the explanatory variables in the equation. But consider a simple nonrecursive model:

$$y_1 = \beta_{12}y_2 + \gamma_{11}x_1 + \zeta_1 \quad (5.2)$$

$$y_2 = \beta_{21}y_1 + \gamma_{22}x_2 + \zeta_2 \quad (5.3)$$

The R^2 for each equation is complicated by the nonrecursive nature of the system. Take Equation 5.2: The problem is that ζ_1 is not uncorrelated with y_2 appearing in that equation. Therefore, we cannot simply divide the variance of y_1 between the disturbance, ζ_1, and the other variables on the right side of this equation (which include y_2). The interpretation of R^2 is therefore unclear. As one solution, Jöreskog (1999) recommends using the R^2 from the reduced-form equation (R^{2*}). The reduced-form R^2 can be interpreted as the relative variance of y explained by all explanatory variables in the system of equations (see also Bentler & Raykov, 2000; Hayduk, 2006; Teel, Bearden, & Sharma, 1986).

Models estimated using IV estimates such as two-stage least squares (2SLS) or three-stage least squares (3SLS) may yield negative R^2 values (software packages may also suppress R^2 values or not show any values). In the context of nonrecursive models, this R^2 value should not be interpreted; as just indicated, researchers are advised to interpret the R^2 value from the reduced-form equation. To understand the negative R^2, consider a naive approach that would compute the residual sum of squares for the model including the instrumented variables $(\hat{y})$, which would always yield a nonnegative R^2. However, we are interested in the structural model including the observed ys, and the residual sum of squares from this model need not be less than the total sum of squares. This does not imply anything wrong about the model, and parameters can still be well estimated. More detail, as well as a small simulation illustrating that a negative R^2 can be obtained even though the population parameters are well estimated by the model, can be found at http://www.stata.com/support/faqs/stat/2sls.html.

Assessing Overall Model Fit

Simultaneous equation models, both recursive and nonrecursive, are a subset of the larger class of SEMs with latent variables (Bollen, 1989b). Therefore, for overidentified models, the full range of goodness-of-fit statistics used in SEM is available for model assessment. Longer treatments are available elsewhere (Bollen, 1989b; Hu & Bentler, 1995, 1999; Kaplan, 2009), so we introduce only a small number of goodness-of-fit statistics here with a focus on the theory of overall model fit for researchers unfamiliar with this type of assessment. These statistics are readily available from SEM software packages, for example, LISREL, MPlus, and AMOS. They are at present more difficult to obtain in other software packages.

Measures of overall fit generally attempt to assess how closely the model-implied covariance matrix at the estimated parameter values approaches the population covariance matrix. Thus, overall model fit again rests on the fundamental statistical hypothesis,

$$\boldsymbol{\Sigma} = \boldsymbol{\Sigma}(\boldsymbol{\theta}) \tag{5.4}$$

The test statistic, sometimes known as the chi-square test, assesses the fundamental statistical hypothesis directly. Its null hypothesis reads

$$H_0\text{: } \boldsymbol{\Sigma} = \boldsymbol{\Sigma}(\boldsymbol{\theta}) \tag{5.5}$$

In other words, our model fits the data perfectly.

Since we only have sample realizations of each of these population covariance matrices, they can differ simply due to sampling fluctuations. The chi-square test accounts for this and has a chi-square distribution with the number of degrees of freedom equal to the number of parameters being estimated: $1/2(p + q)(p + q + 1)$. The chi-square test statistic, $T = F_{\text{ML}}(N - 1)$, tests whether $\boldsymbol{\Sigma}$ and $\boldsymbol{\Sigma}(\boldsymbol{\theta})$ are equal, using the sample realizations $\mathbf{S}$ and $\boldsymbol{\Sigma}(\hat{\boldsymbol{\theta}})$. The larger the discrepancy between $\mathbf{S}$ and $\boldsymbol{\Sigma}(\hat{\boldsymbol{\theta}})$, the greater the value of F_{ML}. Rejection of the null hypothesis (a significant χ^2 test) suggests some problem with the model's fit to the data. A nonsignificant χ^2 test suggests good model fit.[2] We cannot perform the test if the model is exactly identified, as there are no degrees of freedom in the model.

Given theoretical uncertainty about models in the social sciences, many researchers consider the chi-square test to be a very restrictive test as it requires exact, or perfect, fit. To reduce reliance on a single indicator of model fit, a number of alternative fit indices have been developed.

---

[2]Assumptions of the chi-square test include (1) the absence of outliers, (2) that the multivariate distribution of the observed variables is normal with respect to kurtosis (Browne, 1984), (3) the sample is sufficiently large for the asymptotic properties of the test, and (4) the sample comes from a single population. Finally, it is important to recognize that with a very high level of statistical power (e.g., very large sample size), the researcher will almost always reject the null hypothesis as long as the model has at least some small misspecification in it. (For more detail, see Anderson & Gerbing, 1984; Curran, West, & Finch, 1996; Tanaka, 1993.)

The incremental fit family of fit statistics relies on the comparison of a baseline model to the researcher's model of interest. Although in principle various possible baseline models could be specified, the most common one assumes zero covariances between the observed variables of the model (thus, only variances are estimated). The chi-square from the baseline model, labeled χ^2_{b}, has degrees of freedom df_{b}. This test statistic is compared with the test statistic from the researcher's specified (maintained) model, denoted χ^2_{m}, with degrees of freedom df_{m}. Fit statistics in the incremental fit family generally range from 0 to 1, with higher values indicating better fit.

The simplest fit statistic in this family is the normed fit index (NFI; Bentler & Bonnett, 1980)

$$\mathrm{NFI} = \frac{\chi^2_{\mathrm{b}} - \chi^2_{\mathrm{m}}}{\chi^2_{\mathrm{b}}} \tag{5.6}$$

which assesses how much the chi-square value is reduced proportionally by moving from the baseline to the maintained model.

The NFI, although easy to understand, is problematic for two reasons. First, it does not account for degrees of freedom and can therefore be improved by adding more parameters to the model. Second, the mean of the sampling distribution of the NFI increases as sample size increases.

An improved measure of fit in this family is the incremental fit index (IFI; Bollen, 1989a), which both takes into account the degrees of freedom in the model and has a mean of the sampling distribution that is not influenced by sample size. Its calculation is as follows:

$$\mathrm{IFI} = \frac{\chi^2_{\mathrm{b}} - \chi^2_{\mathrm{m}}}{\chi^2_{\mathrm{b}} - df_{\mathrm{m}}} \tag{5.7}$$

To understand the IFI, note that the expected value of the chi-square of a correct maintained model is df_{m}, making the denominator the standard with which the numerator is compared. That is, the numerator will be less than the denominator if your model is incorrect (because the chi-square for the maintained model will be larger than its degrees of freedom). So higher values of the IFI indicate better fit.

Two other fit measures in this family are the Tucker-Lewis Index (TLI; Tucker & Lewis, 1973) and the comparative fit index (CFI; Bentler, 1990). Each again computes the difference between the baseline model and the maintained model adjusting for degrees of freedom. For example, TLI is

$$\text{TLI} = \frac{(\chi^2_{\text{b}}/df_{\text{b}}) - (\chi^2_{\text{m}}/df_{\text{m}})}{(\chi^2_{\text{b}}/df_{\text{b}}) - 1} \tag{5.8}$$

A second family of fit measures includes the goodness-of-fit index (GFI) and the adjusted goodness-of-fit index (AGFI; Jöreskog & Sörbom, 1986). The GFI is calculated as

$$\text{GFI} = 1 - \frac{\text{tr}\left[\left(\boldsymbol{\Sigma}^{-1}(\hat{\boldsymbol{\theta}})\mathbf{S} - \mathbf{I}\right)^2\right]}{\text{tr}\left[\left(\boldsymbol{\Sigma}^{-1}(\hat{\boldsymbol{\theta}})\mathbf{S}\right)^2\right]} \tag{5.9}$$

where $\boldsymbol{\Sigma}^{-1}(\hat{\boldsymbol{\theta}})$ is the inverse of the implied covariance matrix at the parameter estimates, **S** is the sample covariance matrix, **I** is the identity matrix, and tr represents the trace of the matrix, or the sum of the diagonal elements. In short, the GFI compares the sample covariances with the model-predicted covariances. The further apart these two matrices, the more is subtracted from 1, suggesting poorer fit. The GFI suffers from the same problems as the NFI—a lack of correction for degrees of freedom and a mean of the sampling distribution that varies with sample size.

The AGFI fixes the first problem by taking into account the model's degrees of freedom. It is calculated as follows:

$$\text{AGFI} = 1 - [(p + q)(p + q + 1)/2df_{\text{m}}]\left[\left(1 - (1 - \frac{\text{tr}\left[\left(\boldsymbol{\Sigma}^{-1}(\hat{\boldsymbol{\theta}})\mathbf{S} - \mathbf{I}\right)^2\right]}{\text{tr}\left[\left(\boldsymbol{\Sigma}^{-1}(\hat{\boldsymbol{\theta}})\mathbf{S}\right)^2\right]}\right)\right] \tag{5.10}$$

Another commonly used measure of goodness of fit is the root mean squared error of approximation (RMSEA; Browne & Cudeck, 1993; Steiger & Lind, 1980), computed as follows:

$$\text{RMSEA} = \sqrt{\frac{(\chi^2_{\text{m}} - df_{\text{m}})/(N - 1)}{df_{\text{m}}}} \tag{5.11}$$

where N is the sample size. Although this measure also ranges from 0 to 1, *smaller* values for this measure indicate better fit.

The computation of the RMSEA is based on the relationship between the test statistic, T, and the noncentral chi-square distribution. When a model is properly specified, T follows a central chi-square distribution with expected value of df. When a model is *not* properly specified, T follows a noncentral chi-square distribution with expected value $df + \lambda$ where λ is the noncentrality parameter. Thus, the noncentrality parameter provides a measure of the degree of misspecification of a hypothesized model. And it is central to the calculation of the RMSEA, which can also be written as

$$\text{RMSEA} = \sqrt{\frac{\hat{\lambda}}{df_{\text{m}}(N-1)}} \tag{5.12}$$

where $\hat{\lambda}$ is the sample estimate of the noncentrality parameter.

The RMSEA is a popular measure of goodness of fit. Part of this popularity arises from a researcher's ability to compute confidence intervals around the RMSEA value. The RMSEA is inflated for properly specified models at small sample sizes (below 200), so researchers with small samples should be aware that the RMSEA may indicate rejection of a well-specified model (Curran, Bollen, Chen, Paxton, & Kirby, 2003). Researchers should therefore use caution when judging models with small samples based on recommended cutoffs for good fitting models (Browne & Cudeck, 1993; Chen, Curran, Bollen, Kirby, & Paxton, 2008). Too many true models are rejected by these cutoffs at small sample sizes.

Overall, if one reports (1 – RMSEA), then all the fit statistics discussed thus far have the same interpretation: The closer the value to 1, the better the model fit. Traditionally, values above .9 for these fit statistics have been considered to represent adequate fit with values above .95 now recognized to suggest excellent fit (Hu & Bentler, 1999). Certainly, these recommended cutoff points for fit measures can be subjective (see Browne & Cudeck, 1993, for the RMSEA specifically), so reporting several measures from different families is a good strategy. Overall, if a model has a nonsignificant chi-square test, with the remaining fit statistics at .95 or above, a researcher can express confidence in the fit of the model to the data. Also common is an ambiguous situation, such as a model with a significant chi-square test and fit statistic values ranging from .9 to .95. In this case, the researcher should express some caution in the fit of the model to the data. In general, we recommend reporting a variety of fit statistics, along with their interpretation, so that readers can judge the adequacy of the model themselves.

The measures described so far provide an assessment of the quality of the fit compared with some general criterion. But it is sometimes useful to compare *between* two or more models. If models are nested—that is, the parameters of one model are a subset of the parameters of the other model—a chi-square difference test can be performed between them. If they are not nested, other approaches are possible. Two information theory measures are useful in this regard. The Akaike Information Criterion (AIC; Akaike, 1974) is calculated as

$$\text{AIC} = \chi^2_{\text{m}} + 2t$$

where t is the number of free parameters. The Bayesian Information Criterion (BIC; Schwarz, 1978) is calculated as

$$\text{BIC} = \chi^2_{\text{m}} + t \log(N)$$

The BIC value can be used as an approximation of the Bayes Factor when comparing several competing models (Raftery, 1995). For each of these measures, when comparing two models, smaller values indicate the preferred model. Both the AIC and the BIC have penalties for more parameters, with the BIC penalizing to a greater extent than the AIC.

Yet another approach that can be used in some instances is a vanishing tetrads test (Bollen & Ting, 1993, 1998, 2000). A tetrad is created from four random variables, and it measures the difference between the product of one pair of covariances and the product of the other pair. Model structures often imply that some population tetrads should be zero, and this is tested with a vanishing tetrads test. The null hypothesis is that $\tau = 0$, where τ is a vector of unique tetrads that should vanish if the model is valid. Rejection of the null hypothesis suggests that the model is incorrect, whereas failing to reject suggests consistency between the model and the data. Determining which tetrads are implied vanishing generally requires use of covariance algebra or empirical strategies using existing software (Hipp, Bauer, & Bollen, 2005). Given that the implied vanishing tetrads can be redundant, it is necessary to determine the unique tetrads, and Hipp and Bollen (2003) suggested a method to accomplish this.

The tetrad test can also assess two models that are nested for vanishing tetrads. This occurs when all implied vanishing tetrads of one model are a subset of the implied vanishing tetrads of another. This difference in the two models is asymptotically distributed as χ^2 with *df* equal to the difference in number of unique vanishing tetrads implied by each model. Although tetrads are not intuitive—nor is understanding whether the models are nested for tetrads—independent software exists for applied researchers to apply these tests (Hipp & Bauer, 2002; Hipp et al., 2005) even in instances in which one or more outcomes are dichotomous or ordinal (Hipp & Bollen, 2003).

We have only described a subset of all possible fit statistics. For a discussion of others, see Bollen (1989b), Kaplan (2009), or any SEM textbook. General consensus is that a researcher should examine and report several of these measures to provide a good overall sense of the quality of the model fit.

Assessing the Quality of the IVs

Since estimation of nonrecursive systems hinges on IVs, it is critical to assess the quality of the IVs used in a model. Remember that we make two key assumptions about IVs: (1) The IVs are not correlated with the disturbance term and (2) the IVs are correlated with the problematic variable (they are reasonably strong predictors of the problematic variable). Assumption (1) is about validity and is assessed with tests of overidentifying restrictions. Assumption (2) is assessed with tests for weak instruments.

Assessing the Validity of the IVs

There is an unfortunate tendency among researchers to focus on IVs solely as a means to identify the model. It is true that finding excluded IVs may allow a model to be identified. However, if the restrictions used to identify the model are not justified, biased and inconsistent estimates are possible. Researchers must therefore assess whether their IVs are valid. Valid IVs are not correlated with the disturbance.

Theory is the place to begin in determining whether an excluded instrument is valid. In a nonrecursive model, researchers need to provide clear and detailed explanations for why an IV influences the problematic variable but is not correlated with the disturbance in the equation of interest. Instruments should not have a direct relationship with the outcome variable in the equation of interest or any omitted variables that influence the outcome variable. In the best possible situation, theory will be reinforced with results from prior research, for example, instruments tested in other populations. If no prior research is available, researchers must make strong a priori theoretical justifications for their instruments. Theory is particularly necessary in situations where a researcher has only a single IV: In this case, an equation is exactly identified and tests of overidentifying restrictions are not available. Chapter 2 of this monograph provides a range of theoretically reasonable IVs found in published research.

If a researcher specifies overidentifying restrictions for a model by excluding more IVs than needed for identification, the appropriateness of the IVs can be assessed. A number of tests of the IVs are available, including the Sargan (1958) test and the Basmann (1960) test, and all are designed to test the assumption that the excluded IVs are uncorrelated with the disturbance term. The idea underlying the tests we discuss is that if the IVs are uncorrelated with the disturbance, they should not explain variance in the 2SLS residuals, meaning that the (uncentered) R^2 from a regression of the 2SLS residuals on the IVs should be zero. Failing a test of overidentifying restrictions suggests that at least one of the excluded IVs correlates with the disturbance.

The Sargan (1958) test is one of the most popular tests of overidentifying restrictions as it is available in Stata and is relatively simple to implement:[3]

$$S = \frac{\hat{\boldsymbol{\zeta}}'\mathbf{Z}(\mathbf{Z}'\mathbf{Z})^{-1}\mathbf{Z}'\hat{\boldsymbol{\zeta}}}{\hat{\boldsymbol{\zeta}}'\hat{\boldsymbol{\zeta}}/N} \tag{5.13}$$

where $\hat{\boldsymbol{\zeta}}$ is the residuals from the 2SLS estimation, $\mathbf{Z}$ is the matrix of all IVs (both included and excluded), and N is the sample size. The null hypothesis is that the instruments are correctly excluded from the equation and are uncorrelated with the error term. The test statistic is computed as NR^2 where the R^2 is uncentered and comes from the regression of the residuals from the 2SLS on the instruments (both included and excluded).[4] The test statistic follows a chi-square distribution with degrees of freedom equal to the degree of overidentification.

In short, to implement the Sargan test, multiply the uncentered R^2 from the regression of the residuals from the 2SLS on the IVs by the sample size (N) and perform a chi-square test of this value. The Sargan test is also available in Stata with the **ivregress** command (Baum, Schaffer, & Stillman, 2003). The null hypothesis is valid instruments, so failure to reject the test statistic is an indication that the instruments may be valid.

Hansen's J test is a more general version of the Sargan test using the generalized method of moments estimator rather than 2SLS (Hansen, 1982). Briefly, the J statistic weights squared deviations of the sample moments evaluated at the generalized method of moments (GMM) estimates. It is consistent in the presence of heteroscedasticity and autocorrelation. When there is conditional homoscedasticity, the Sargan statistic and

---

[3]See also Anderson and Rubin (1949) and Hausman (1978, 1983).

[4]The usual definition of R^2 uses the total sum of squares in deviation from the sample mean and can be considered the variation in y after fitting the constant. The uncentered R^2 instead uses the total sum of squares of y around 0.

Hansen's *J* test are equivalent. In the presence of heteroscedasticity, Hansen's *J* statistic remains consistent, whereas the Sargan statistic is not.

The *J* statistic has a chi-square distribution with degrees of freedom equal to the number of overidentifying restrictions. Like the Sargan test, this test is also implemented in Stata using the **estat overid** postestimation command after **ivregress**.

Another test of overidentifying restrictions is the Basmann (1960) test, which is similar in principle to the Sargan test:

$$B = \frac{(\hat{\zeta}'\mathbf{Z}(\mathbf{Z}'\mathbf{Z})^{-1}\mathbf{Z}'\hat{\zeta}) / (L-K)}{(\hat{\zeta}'(I-\mathbf{Z}(\mathbf{Z}'\mathbf{Z})^{-1}\mathbf{Z}')\hat{\zeta}) / (N-L)} \tag{5.14}$$

where *L* is the total number of instruments, *K* is the number of endogenous variables on the right-hand side, and *N* is the sample size. The Basmann test statistic is distributed as an *F* statistic ($df = L - K$ and $N - L$) and is asymptotically equivalent to the Sargan test (Baum et al., 2003). A significant *F* test indicates that at least one of the IVs is correlated with the disturbance. To implement this test in SAS, add "overid" as an option to the model statement. In Stata, the Basmann test is available with **ivregress** postestimation commands.

Again, these tests require that the equation is overidentified: that there are more excluded instruments than problematic variables in the equation. In the case of nonrecursive models, then, a researcher must have more excluded instruments than endogenous variables in the equation.

More important, these tests assume at least one valid instrument. Thus, although these tests can be used to assess instrument validity, it is ultimately untestable. Researchers will always rely partially on theory that they have at least one valid instrument.

Researchers are encouraged to report the results of both the Basmann and Sargan tests, as Monte Carlo simulation studies differ as to which has the most desirable features and best performance in finite samples (Kirby & Bollen, 2009; Magdalinos & Symeonides, 1996). At small sample sizes (e.g., below 100), both Sargan's test and the Basmann test reject too many correctly specified equations, with Sargan's test performing slightly better (Kirby & Bollen, 2009).

While these tests can be viewed as a test of a correlation of the instruments with the disturbance, they can also be viewed as a test of model specification. This is because if an equation fails an overidentifying restriction test, it may not be the equation itself that is the cause of misspecification. Instead, the misspecification could be located in another equation that would change the list of excluded instruments for the equation of interest. In short, while these tests can be viewed as testing a specific property of IVs—whether they are valid—they can also be viewed as overall tests of model specification.

Testing the Strength of the Instruments

The second key feature of an IV is that it correlates with its endogenous, or problematic, variable. While violating the first assumption—that the IVs are uncorrelated with the disturbance—can lead to inconsistent estimates, violating the second assumption will lead to inefficient estimates and increased finite sample bias.[5] This is not solely a problem for limited-information methods; weak instruments can bias estimated parameters and distort hypothesis testing regardless of the estimation procedure employed. Assessing the strength of IVs as predictors is therefore an important feature of nonrecursive models (Bound, Jaeger, & Baker, 1995).

In gauging the strength of the IVs, we can consider the amount of variance explained in the first-stage regressions (R^2), where the first stage is the regressions of the endogenous variables on the full set of instruments (included and excluded). Consider an instance in which very little variance is explained in the first-stage regression. In this case, the predicted value being substituted in the second-stage equation is mostly white noise. Researchers using latent variable SEM software programs are frequently unaware of the explanatory power of the instruments, since programs generally do not provide information about the first-stage regressions. Inspection of these R^2 values provides useful information regarding the IVs.

The key issue in assessment is actually the partial variance the identifying (excluded) IVs explain in the first-stage equation, beyond the variance explained by the full set of instruments. This is the partial R^2. To understand its importance, consider an example in which we naively estimate a model with no excluded IVs. If the R^2 of the first-stage equation is, say, .50, then the R^2 will also be .50 when estimating the first-stage equation without the excluded IVs, since there are none in this example. This lack of difference in the R^2s is important. We obtain predicted values from the full reduced-form equation and then estimate the second-stage equation with these estimated values in place of the endogenous variable. The predicted values in this case are simply a linear combination of the other variables in the

[5]A weak instrument can cause finite sample bias in any estimator. Bias can still occur in large samples if instruments are particularly weak. Further, in a model with weak instruments, any correlation, even small, with the error, can lead to inconsistency (Bound, Jaeger, & Baker, 1995).

structural equation. There is perfect multicollinearity between this predicted value and the other variables in the structural equation.

Now consider a situation where we model an additional (excluded) IV that has very little effect on the endogenous variable in the first-stage equation. When we estimate the first-stage equation including all the included IVs and this new IV, we would obtain a relatively similar R^2 value, say .5002. This very small incremental R^2 (.5002 – .50) indicates that the predicted value from the first-stage equation is *nearly* a linear combination of the other variables in the structural equation. When the IVs have very little effect, this predicted value will be highly correlated with the other predictors in the second-stage equation, inducing a high level of collinearity. The predicted variable will likely have a high standard error due to this inefficiency in the second-stage equation.

If, on the other hand, there is an excluded IV (or set of IVs) that explains a reasonable amount of the partial variance in this first-stage equation, the first-stage equation including this IV might have an R^2 of .60. This implies an incremental R^2 of .10 and indicates that this predicted value is considerably different from what it would have been if the IV(s) was left out of the equation. This situation suggests a quality IV. It is therefore important that the variance explained in the first-stage regression improve considerably when adding the excluded instruments.[6]

Researchers including Bound et al. (1995) and Staiger and Stock (1997) present a test of the predictive strength of the IVs that uses the notion of the partial R^2 and its associated F statistic. They outline the following steps:

1. Estimate the reduced-form equation of interest without the excluded IVs,

$$y_1 = \Pi_1^* \mathbf{X} + \zeta_1^{**}$$

where $\mathbf{X}$ represents the included IVs.

2. Estimate the reduced-form equation of interest including the excluded IVs that we represent as $\mathbf{Z}$. That is, estimate the reduced form using the full set of instruments

$$y_1 = \Pi_1 \mathbf{X} + \Pi_2 \mathbf{Z} + \zeta_1^*$$

3. Obtain the partial R^2 and the associated F statistic.

---

[6]The researcher should entertain the possibility that some of these relationships may be nonlinear (or polynomial) rather than relying solely on linear specifications of exogenous variables (Kelejian, 1971).

This F test can help us gauge the finite-sample bias of 2SLS relative to OLS for a nonrecursive model. As Bound et al. (1995) point out, F statistics close to 1 indicate cause for concern. Indeed, Staiger and Stock (1997) suggest that instruments are weak if the first-stage F statistic for the excluded instruments is less than 10 (but see more extended recommendation below). Inspecting the R^2 values for each of these equations is also informative.

When an equation contains multiple endogenous regressors, the partial R^2 measure must also account for the intercorrelations among the instruments (Shea, 1997). In either case, these statistics are available in Stata using the **estat firststage** postestimation command associated with ivregress.

How should one assess the partial R^2 and its related F statistic? As already noted, an early rule of thumb was that F-statistic values above 10 were needed. Stock and Yogo (2005) have extended this assessment by providing critical values on which to focus. Stock and Yogo provide critical values for testing two null hypotheses: (1) that the bias of 2SLS relative to OLS exceeds certain amounts (10%, 15%, 20%, or 30% of the bias of OLS) and (2) that the true significance level of hypothesis tests using 2SLS will be below certain values (10%, 15%, 20%, 25%) when the reported nominal level from 2SLS is actually 5%. Both are hypotheses of weak instruments. The first hypothesis deals with the level of bias in 2SLS, and the second hypothesis deals with issues of inference based on the appropriateness of the size of the standard errors from the analysis. Rejection suggests that the 2SLS estimates are not overly biased and that inference is valid. A portion of the Stock and Yogo critical values table is reproduced as Table 5.1. Readers are referred to Stock and Yogo (2005) for the complete tables.[7]

Testing Endogeneity

The discussion above reminds us that although estimators taking into account potential endogeneity of the model (i.e., 2SLS, 3SLS, full-information maximum likelihood [ML]) are consistent, there is an increase in variance that occurs with the use of estimators accounting for the nonrecursive nature of the model. Thus, researchers may want to test whether there is really endogeneity in their model necessitating the use of the more complicated model specification rather than OLS.

The Hausman endogenity test provides a method for assessing whether a possibly endogenous variable is truly endogenous. To illustrate the test,

---

[7]If one has multiple endogenous variables, the Cragg-Donald statistic should be used rather than the F statistic. This statistic is also available in Stata.

Table 5.1 Selected Critical Values for One Endogenous Regressor From Stock and Yogo (2005)

Number of Instrumental Variables	*Maximal Bias of 2SLS Relative to OLS*				*Actual Significance Level When Nominal Level Is 5%*			
	0.05	*0.10*	*0.20*	*0.30*	*0.10*	*0.15*	*0.20*	*0.25*
1	—	—	—	—	16.38	8.96	6.66	5.53
2	—	—	—	—	19.93	11.59	8.75	7.25
3	13.91	9.08	6.46	5.39	22.30	12.83	9.54	7.80
4	16.85	10.27	6.71	5.34	24.58	13.96	10.26	8.31
5	18.37	10.83	6.77	5.25	26.87	15.09	10.98	8.84
10	20.74	11.49	6.61	4.86	38.54	20.88	14.78	11.65
15	21.23	11.51	6.42	4.63	50.39	26.80	18.72	14.60
20	21.38	11.45	6.28	4.48	62.30	32.77	22.70	17.60
30	21.42	11.32	6.09	4.29	86.17	44.78	30.72	23.65

SOURCE: Values taken from Stock and Yogo (2005, Tables 5.1 and 5.2, pp. 100–101). For critical values in the case of more than one endogenous regressor, see Stock and Yogo (2005).

NOTE: 2SLS = two-stage least squares; OLS = ordinary least squares.

return to the simple nonrecursive system outlined in Equations 5.2 and 5.3 above. The first equation in that two-equation system is

$$y_1 = \beta_{12}y_2 + \gamma_{11}x_1 + \zeta_1 \tag{5.15}$$

where y_2 is endogenous and x_2 is an excluded IV for this equation.

The Hausman test is derived by regressing the potentially endogenous variable, y_2, on all instruments (included and excluded),

$$y_2 = \Pi_{21}x_1 + \Pi_{22}x_2 + \zeta_2^* \tag{5.16}$$

and saving the residuals $\hat{\zeta}_2$.

If y_2 is exogenous, then those residuals, $\hat{\zeta}_2$, should be uncorrelated with the original structural error, ζ_1, which we can test by regressing ζ_1 on $\hat{\zeta}_2$:

$$\zeta_1 = \rho_1\hat{\zeta}_2 + e_1 \tag{5.17}$$

where y_2 is exogenous if $\rho_1 = 0$.

In practice, we substitute 5.17 into 5.15 to obtain

$$y_1 = \beta_{12}y_2 + \gamma_{11}x_1 + \rho_1\hat{\zeta}_2 + \text{error} \quad (5.18)$$

and a *t*-test for ρ_1 is a test of the null hypothesis that $\rho_1 = 0$, or that y_2 is exogenous. If the null hypothesis is rejected, it is evidence that y_2 is endogenous, and 2SLS should be used.[8]

The validity of the Hausman test rests on the appropriateness of model specification. The endogeneity test performs poorly in the presence of weak instruments (Hahn & Hausman, 2002; Jeong & Yoon, 2010; Staiger & Stock, 1997). Thus, the researcher is confronted with the ironic challenge that for one to conclude that endogeneity is not present requires accurately specifying the proper model under the assumption of endogeneity. Failing to assess the quality of the instruments can lead to an inappropriate conclusion.[9]

Assessment in Practice: The Example of Trust and Association Memberships

We continue the example of the reciprocal relationship between interpersonal trust and voluntary associations to apply principles of assessment outlined in this chapter. We begin with a discussion of validity and move to assessment of the strength/weakness of the IVs and to overall assessment. Briefly, recall that trust and civic engagement are modeled in a reciprocal relationship. One model is exactly identified with one IV for each of our endogenous variables and one measure (education) predicting both of these outcomes. The second is overidentified with two IVs for each endogenous variable and one measure predicting both of the outcomes.

The first model in Table 5.2 presents results of the exactly identified model estimated with 2SLS. With just a single identifying IV, we are unable

---

[8]The original reasoning behind the Hausman test relies on the comparison between the OLS and 2SLS estimators as a test of endogeneity. In brief, if the supposedly endogenous variable in an equation is in fact exogenous, then OLS and 2SLS should differ only by sampling error (see Wooldridge, 2002, or Baum et al., 2003, for more information). The test presented here may be more intuitive. Note also that this is a pretest estimator, in that the estimator is chosen based on the outcome of this pretest. The subsequent estimated model is not an initial test of the data, therefore, and raises issues of multiple testing and the need to adjust parameter tests (see Guggenberger, 2010).

[9]One recent approach proposes estimating models with latent instrumental variables (Ebbes, Wedel, Böckenholt, & Steerneman, 2005).

Table 5.2 Results for Exactly and Overidentified Models, 2SLS

	Exactly Identified Model	*Overidentified Model*
Association memberships outcome		
Interpersonal trust	0.027	0.076
	(0.309)	(0.275)
Years of education	0.184	0.168
	(0.021)	(0.018)
Hours of TV viewing	—	–0.066
		(0.014)
Presence of children less than 6 years of age	–0.189	–0.178
	(0.060)	(0.056)
Intercept	–0.405	–0.002
	(0.284)	(0.230)
R^2	0.108	0.117
R^2 from reduced-form equation	0.070	0.077
Partial R^2	0.010	0.012
F statistic	45.977	28.911
Sargan test	—	0.538
Basman test	—	0.538
Interpersonal trust outcome		
Association memberships	0.679	0.474
	(0.179)	(0.090)
Years of education	–0.064	–0.025
	(0.034)	(0.017)
Experienced a burglary in the past year	–0.300	–0.298
	(0.075)	(0.060)
Parents divorced at age 16	—	–0.072
		(0.032)
Intercept	–0.531	–0.605
	(0.118)	(0.080)
R^2	—	—
R^2 from reduced-form equation	0.106	0.112
Partial R^2	0.004	0.011
F Statistic	19.580	25.456
Sargan test	—	2.150
Basmann test	—	2.149

NOTE: N = 4,598; 2SLS = two-stage least squares.

to test the validity of the IVs; thus, we must rely on the theory presented in Chapter 2 to justify our choice of instruments. We will, however, be able to assess the strength of the instruments as well as perform test of endogeneity. The second model in Table 5.2 presents the model results including two IVs for each equation. This specification is overidentified and allows us to test the validity of the IVs as well as their strength.

To estimate the overidentified model in Stata 11, we use the following code:

```
*voluntary associations equation
ivregress 2sls assoc (intprtrst = burglary pardiv16) educ tvhours babies, first
*interpersonal trust equation
ivregress 2sls intprtrst (assoc = babies tvhours) educ burglary pardiv16, first
```

The first line of the code estimates the structural equation with the voluntary associations variable as the outcome using the **ivregress** command. The **2sls** command tells Stata to use this estimator. The second line of the code estimates the structural equation with the interpersonal trust variable as the outcome. The option after the comma (**first**) asks Stata to provide the first-stage equation results.

There are several postestimation (**estat**) commands that can be used after estimating the equation to obtain additional information. The command **estat firststage** reports the first-stage regression statistics. If there is a single endogenous regressor, it reports the regression results from the first-stage equation, the partial R^2 and the F test for the exclusion restrictions for the excluded IVs. It also reports relevant lines from the Stock and Yogo critical values tables for tests of the weakness of the instruments. If there is more than one endogenous regressor, it reports Shea's partial R^2 and adjusted R^2 and reports the tests of the weakness of the instruments.

The command **estat overid** reports the Basmann (1960) and Sargan (1958) chi-square tests of overidentifying restrictions.

The command **estat endogenous** reports two Hausman-based endogeneity tests: the Durbin (1954) chi-square test of endogeneity and the Wu-Hausman F test of endogeneity (Hausman, 1978; Wu, 1974). The Wu-Hausman F test produces the Hausman test as described above generalized to the case of multiple endogenous variables. The Durbin test uses an alternative estimate of the error variance and is more efficient (see Baum et al., 2003).

Returning to the output in Table 5.2, the R^2 from these first-stage equations can be used as a broad indicator of the goodness of fit of the model. In the example, the R^2s from the first-stage equations for the overidentified models with two instruments for each equation are .112 for the interpersonal trust equation and .077 for the equation with voluntary associations as an outcome. These values provide a sense of the extent to which the instrumented variable is being explained in the first-stage equation. The key R^2 to consider in the first stage is the partial R^2, however, which we discuss below.

Assessing the Validity of the IVs

In the overidentified example, the instruments for the interpersonal trust equation (the presence of children less than 6 years of age and the number of hours watching TV) appear reasonable based on the tests introduced above: The chi-square value of 2.15 with 1 degree of freedom ($p = .14$) for the Sargan test suggests that we are unable to reject the null hypothesis that these instruments are indeed valid. The Basmann test yields an identical result. For the voluntary associations equation the chi-square of .538 with 1 degree of freedom ($p = .46$) also suggests that we should not reject the null hypothesis that the instruments are valid. The Basmann test returns an identical result. Remember, however, that these tests assume at least one valid instrument.

What if a clearly improper IV is employed? For instance, rather than using the measure of being robbed as an instrument, imagine instead a household's income is included. Although a household's income arguably affects interpersonal trust (we found a positive coefficient for this measure in our first-stage equation), there is some theoretical reason to believe that it also affects association memberships. The Basmann and Sargan tests detect this problem (with chi-squares of 22.7 and 22.8 on 1 *df*), indicating clear invalidity of at least one IV. Results of this reestimated model specification are shown in the second column of Table 5.3. Note that the effect of interpersonal trust now appears as a very strong positive effect in this misspecified equation. Furthermore, the estimates of the other coefficients in the equation have also been affected, some of them considerably. This exercise highlights the importance of assessing the appropriateness of IVs.[10]

To address heteroscedasticity and/or autocorrelation, we could request the GMM estimator in Stata, which provides the Hansen J test. In the

---

[10]See Young (2009) and Bound et al. (1995) for evaluations of published research using IVs.

Table 5.3 Comparing Results With an Inappropriate IV, Overidentified Model, 2SLS

	Using Burglarized and Divorced Parents as IVs	*Using Burglarized and Real Income as IVs*
Association memberships outcome		
Interpersonal trust	0.076	1.164
	(0.275)	(0.241)
Years of education	0.168	0.106
	(0.018)	(0.017)
Hours of TV viewing	−0.066	−0.037
	(0.014)	(0.015)
Presence of children	−0.178	−0.035
less than 6 years of age	(0.056)	(0.057)
Intercept	−0.002	0.760
	(0.230)	(0.218)
Sargan test	0.538	22.747
Basmann test	0.538	22.830

NOTE: $N = 4{,}598$; IV = instrumental variable; 2SLS = two-stage least squares.

overidentified example, Hansen's J statistic is almost identical to Sargan/Basmann at .55 ($p = .46$) for the associations equation and 2.34 ($p = .13$) for the trust equation.

The Strength of the IVs

Strength of the IVs can be assessed by estimating the reduced-form equation with and without the excluded instruments (Bound et al., 1995; Staiger & Stock, 1997). For instance, in the equation for voluntary associations, we would estimate the first-stage equation in which interpersonal trust is regressed on education, TV hours, and the presence of young children (thus, none of the excluded instruments are included). For the model in which we included a single identifying instrument, this equation would be reestimated with the burglary variable added as a predictor: The partial R^2 of this instrument is .011. Performing an F test of the difference in these two models, as outlined by Staiger and Stock (1997), indicates that this identifying instrument explains a large enough portion of the variance in this first-stage equation, as the F statistic of 45.98 ($df = 1, 4594$) shows a highly significant improvement ($p < .0001$). For the model with two identifying instruments, the partial R^2 of these instruments is .012, with an F test

(28.91, *df* = 2, 4592, $p < .0001$) suggesting that we do not have a weak instruments problem.

We need to compare these *F* statistics to the Stock and Yogo critical values. Assessment of relative bias is not possible with fewer than three IVs. But assessment of the true significance level of the hypothesis tests can be undertaken. For the just-identified (one-instrument) equation, if we are willing to accept only a 10% rejection rate (or below) when the nominal Wald test is 5%, then the critical value is 16.38.[11] With a maintained *F* statistic of 45.98, we therefore reject the null hypothesis of weak instruments. Similarly, for the overidentified (two-instrument) model, we obtain a critical value of 19.93. With an *F* statistic of 28.91, we therefore reject the null hypothesis of weak instruments.

In the interpersonal trust equation with one identifying instrument, the partial R^2 is .004, and the *F* statistic (19.58, *df* = 1, 4594, $p < .0001$) indicates a significant improvement in the fit of this first-stage equation. For the equation with two identifying instruments, the partial R^2 is .01 with a significant associated *F* test (25.46, *df* = 2, 4594, $p < .0001$). Again, with only one or two instruments, assessment of relative bias is not possible. As before, for the just-identified trust equation, if we are willing to accept only a 10% rejection rate when the nominal Wald test is 5%, then the critical value is 16.38. The *F* statistic of 19.58 does pass this critical value. For the overidentified trust equation the *F* statistic of 25.46 also passes the critical value of 19.93. In the presence of weak instruments, researchers should consider alternative methods for estimation and testing (Anderson & Rubin, 1949; Andrews, Moreira, & Stock, 2006; Andrews & Stock, 2007; Fuller, 1977; Hahn, Hausman, & Kuersteiner, 2004; Moreira, 2003; see Murray, 2006a, for a partial review).

Assessing Endogeneity

We can assess whether 2SLS is necessary by performing the Hausman-based tests of endogeneity. Stata provides two flavors of the Hausman test and both reach the same conclusion. In the overidentified interpersonal trust equation, both the Durbin chi-square test (41.57, $p = .000$) and the Wu-Hausman *F* test (41.90, $p = .000$) reject the null hypothesis that the variables are exogenous. In contrast, in the overidentified voluntary associations equation, both tests are nonsignificant (Durbin: .14, $p = .70$; Wu-Hausman: .14, $p = .70$), suggesting that variables may be treated as exogenous. As a pretest estimator, these results need to be interpreted with caution (Guggenberger, 2010).

---

[11]See Line 1 of Table 5.1. Also, Stata reports the relevant critical values.

Assessing Overall Model Fit

For the overidentified model, we can assess overall model fit using the range of goodness-of-fit statistics available in the SEM with latent variables tradition. Significant chi-square statistics reject perfect fit between data and model and are, therefore, taken as an indicator of poor model fit. Nonsignificant chi-square statistics are an indication of good fit. Our overidentified example model has a chi-square test value of 2.56 with 2 *df* ($p = .28$). The nonsignificant value suggests good model fit. The closer the IFI and other incremental measures of fit to 1.0, the better the fit of a model. Typically, values above .90 are considered acceptable and .95 considered optimal. The example model has an IFI of 1.0, suggesting excellent model fit. The other measures of fit in the incremental fit family bolster this claim: the NFI is 1.0 and the TLI is .998. The GFI and AGFI, at .999 and .998, respectively, also suggest excellent model fit. RMSEA values below .05 are typically considered to indicate optimal fit (Browne & Cudeck, 1993). The RMSEA point estimate for the trust/associations model is .008, with a 90% confidence interval (0–.03) falling entirely below .05 as well.

Summary and Best Practices

For nonrecursive models, researchers should focus on three areas: (1) assessing the component fit of each individual equation, (2) assessing overall model fit, and (3) assessing the quality of the IVs. Assumptions regarding IVs in nonrecursive models have important consequences. Violating the first assumption—that the IVs are uncorrelated with the disturbance—can lead to inconsistent estimates. Violating the second assumption regarding the correlation of the IVs with the endogenous variable will lead to inefficient estimates and increased finite sample bias.

A nonrecursive model is therefore only as good as its IVs, and gauging the quality of these instruments is an important, although often overlooked, part of these models. Pragmatically and practically, researchers are advised to proceed with great care in selecting IVs and including them in empirical models. To begin, the importance of a strong theoretical rationale for the excluded IV(s) cannot be overstated. It is possible to find support for an instrument based on studies from other populations, though this should be done with caution. Parsimony is advised, and researchers should generally attempt to use fewer rather than more excluded IVs. Note, however, that with a large sample, researchers can gain efficiency from multiple IVs, provided those IVs are valid. Regardless, searching for variables that are empirically unrelated (i.e., not correlated) to the outcome variable is not recommended.

Researchers should carefully weigh the costs and benefits of the number of IVs included in a model. For instance, if there are instruments available that cannot be included in the same model, it may be useful to estimate separate models using each of the instruments and comparing the results (as an example of this approach, see Hoxby, 1996). Broadly similar results using different instruments may lend confidence to the findings. On the other hand, if the results differ appreciably, this may raise concerns about the quality of at least one of the instruments.

In this chapter, we stressed the importance of a strong theoretical rationale for IVs in addition to assessing the quality of IVs through employing a variety of empirical tests. Finally, although much research is focused primarily on testing the researcher's model of interest, another possible theoretical and empirical exercise in social science research is testing competing models. Such critical tests of competing models are a useful exercise too infrequently conducted.

CHAPTER 6. MODEL INTERPRETATION

In this chapter, we discuss how to interpret model results. Understanding the effect of one variable on another is more complicated in simultaneous equations models than in linear regression models. Rather than focusing solely on direct effects, simultaneous equations models often imply indirect effects. To review, direct effects are effects from one variable to another variable that are not mediated by any other variable in the model. Indirect effects are paths from one variable to another that travel through at least one other variable. Further complications arise in the case of nonrecursive systems of equations, where feedback effects must be addressed.

In this chapter, we show how to calculate and assess effects in recursive and nonrecursive simultaneous equations models. We begin by introducing causal effects and noncausal associations in a discussion of the total association between variables with specific attention to multiplier effects in nonrecursive systems of equations. We then show how to calculate indirect and total effects in both noncomplicated and complicated models. After a brief discussion of mediation and its relationship to indirect effects, we discuss a number of ways to calculate standard errors for indirect effects. The chapter ends with suggestions for programming standard errors of indirect effects.

The Total Association Between Variables: Causal Effects and Noncausal Associations

The total association between any two variables can be divided into two categories: causal effects and noncausal associations. Although researchers are most interested in interpreting a model's causal effects—the direct, indirect, or total effects between variables—these represent only part of the total association between two variables. Noncausal associations also contribute to the total association between two variables and take the form of covariance due either to some unanalyzed association among exogenous variables or to the joint dependence of two variables on a single variable or correlated variables.

To understand total association, causal effects, and noncausal associations, it is useful to turn again to covariance algebra. Consider the recursive model shown in Figure 6.1.

Figure 6.1 Simple Recursive Model

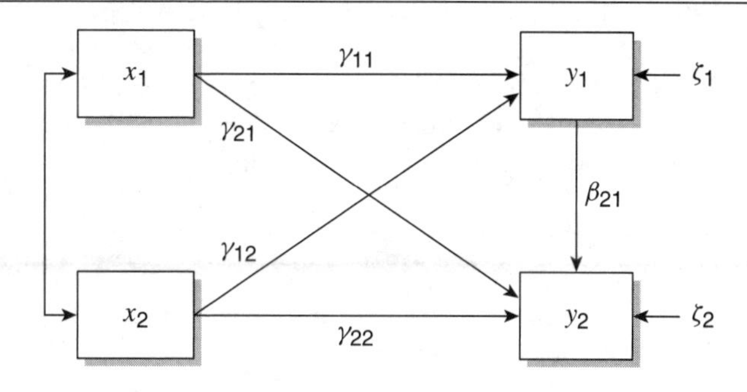

Covariance algebra helps us understand the total association between x_1 and y_1:

$$\mathrm{COV}(x_1, y_1) = \gamma_{11}\mathrm{VAR}(x_1) + \gamma_{12}\mathrm{COV}(x_1, x_2) \tag{6.1}$$

Part of the association between x_1 and y_1 is causal due to the direct effect γ_{11}. Part of it is noncausal, due to x_1's association with another cause x_2, as shown in bold in Figure 6.2.

Figure 6.2 Noncausal Association Between x_1 and y_1

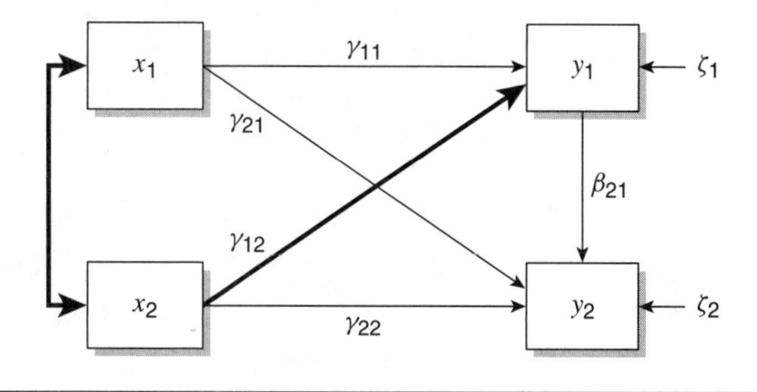

Consider next the covariance between x_1 and y_2:

$$\begin{aligned}\mathrm{COV}(x_1, y_2) = {} & \beta_{21}\gamma_{11}\mathrm{VAR}(x_1) + \beta_{21}\gamma_{12}\mathrm{COV}(x_1, x_2) + \\ & \gamma_{21}\mathrm{VAR}(x_1) + \gamma_{22}\mathrm{COV}(x_1, x_2)\end{aligned} \tag{6.2}$$

There are four components to this covariance: (1) a direct effect from x_1 to y_2 (γ_{21}VAR(x_1)), (2) an indirect effect of x_1 on y_2 through y_1 ($\beta_{21}\gamma_{11}$VAR(x_1)), (3) a noncausal association due to x_1's covariance with x_2 and x_2's direct effect on y_1 (γ_{22}COV(x_1, x_2)), and (4) a noncausal association due to x_1's covariance with x_2 and x_2's indirect effect on y_2 ($\beta_{21}\gamma_{12}$COV(x_1, x_2)). The last noncausal association is illustrated in bold in Figure 6.3.

Figure 6.3 A Noncausal Association Between x_1 and y_2

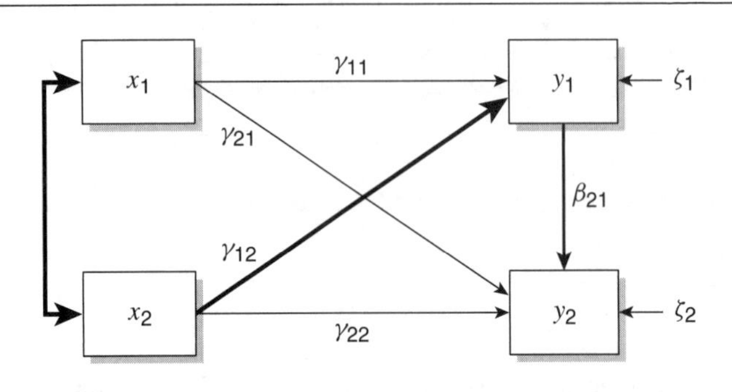

Understanding the total association between two endogenous variables (in this example y_1 and y_2) can be similarly calculated. In this case, there are two additional noncausal ways that y_1 and y_2 can be associated. First, part of the association between y_1 and y_2 is due to the presence of common causes, specifically the effects of x_1 ($\gamma_{11}\gamma_{21}$VAR(x_1)) and of x_2 ($\gamma_{12}\gamma_{22}$VAR(x_2)) on each. The covariance between y_1 and y_2 due to the common cause x_1 is illustrated in Figure 6.4.

Figure 6.4 A Noncausal Association Between y_1 and y_2 Due to Common Cause x_1

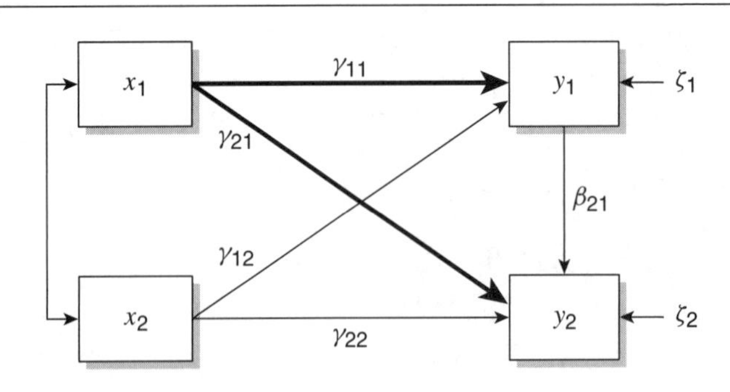

Another part of the total association between y_1 and y_2 is due to a common dependence on correlated causes. The two exogenous variables x_1 and x_2 each affect y_2 and y_1, and the correlation between them induces additional noncausal associations between the two endogenous variables: $\gamma_{12}\gamma_{21}\mathrm{COV}(x_1, x_2)$ and $\gamma_{11}\gamma_{22}\mathrm{COV}(x_1, x_2)$. Figure 6.5 illustrates the former in a path diagram.

Figure 6.5 A Noncausal Association Between y_1 and y_2 Due to Correlated Causes

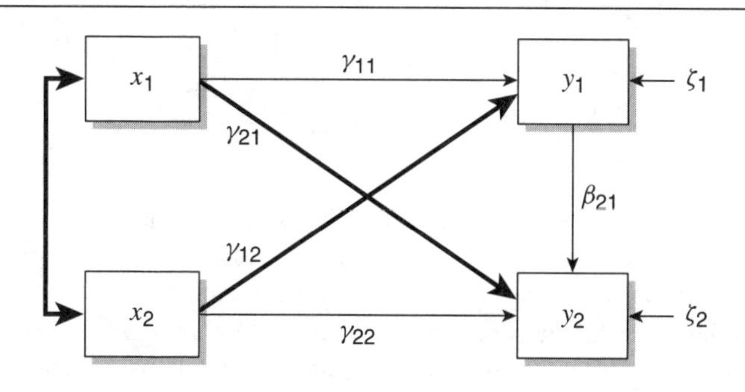

The total effect of one variable on another is therefore not the same as the total association between two variables. The total effect refers to the posited causal effects and is composed of direct effects and indirect effects. The total association between two variables is composed of both these causal effects but also any noncausal associations in the model.

Understanding Multiplier Effects in Nonrecursive Models

For nonrecursive models, understanding the effect of one variable on another can be complicated by the need to account for reciprocal relationships between endogenous variables. Consider the nonrecursive model shown in Figure 6.6.

Covariance algebra provides the association between x_1 and y_1:

$$\begin{aligned}\mathrm{COV}(x_1, y_1) &= \frac{1}{1-\beta_{12}\beta_{21}}[\gamma_{11}\mathrm{VAR}(x_1) + \beta_{12}\gamma_{22}\mathrm{COV}(x_1, x_2)] \\ &= \frac{1}{1-\beta_{12}\beta_{21}}[\gamma_{11}\mathrm{VAR}(x_1)] \\ &\quad + \frac{1}{1-\beta_{12}\beta_{21}}[\beta_{12}\gamma_{22}\mathrm{COV}(x_1, x_2)]\end{aligned} \tag{6.3}$$

Figure 6.6 A Simple Nonrecursive Model

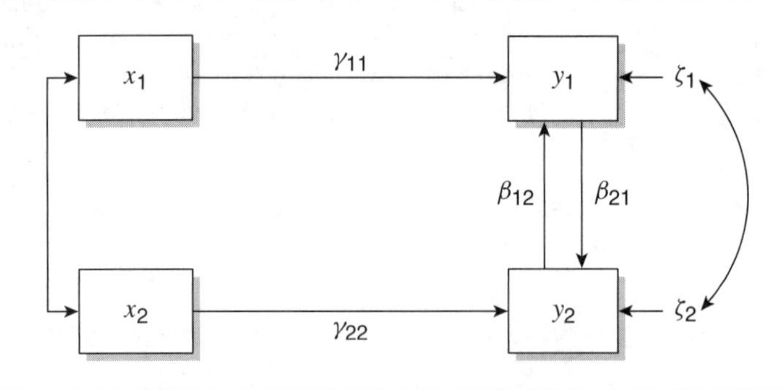

Looking at the covariance between x_1 and y_1, we immediately see that each path is multiplied by $1/(1 - \beta_{12}\beta_{21})$. This is a multiplier that accounts for the reciprocal relationship between y_1 and y_2.

To understand the multiplier, consider that in a nonrecursive model such as that in Figure 6.6, a unit increase in y_1 will ultimately have a return effect on itself. For example, if we assume that $\beta_{21} = .6$ and $\beta_{12} = .5$, then a one-unit increase in y_1 will cause a (1)(.6)(.5) return effect of .3 on y_1. Now y_1 is increasing by .3, and the increase will travel around the loop, resulting in a (.3)(.6)(.5) return effect of .09. That .09 then travels around the loop, and so on, creating smaller and smaller changes in y_1.

Now consider the more typical effect of a change in y_1 on y_2. If y_1 increases one unit, the effect on y_2 is

$$\Delta y_2 = \beta_{21} + \beta_{21}(\beta_{12}\beta_{21}) + \beta_{21}(\beta_{12}\beta_{21})^2 + \beta_{21}(\beta_{12}\beta_{21})^3 + \cdots \quad (6.4)$$

which as a convergence value of an infinite series is

$$= \frac{\beta_{21}}{1 - \beta_{12}\beta_{21}} \text{ if } |\beta_{12}\beta_{21}| < 1 \quad (6.5)$$

So the total effect of y_1 on y_2 is $\beta_{21}/(1 - \beta_{12}\beta_{21})$.

Again, $1/(1 - \beta_{12}\beta_{21})$ can be considered the multiplier, in that any effect on the reciprocal variables must be adjusted by the multiplier.

Returning to the covariance between x_1 and y_1 in Equation 6.3 above, the association between x_1 and y_1 comprises two pieces: the direct effect $\gamma_{11}\text{VAR}(x_1)$ multiplied by the multiplier, $1/(1 - \beta_{12}\beta_{21})$, and a noncausal effect due to x_1's covariance with x_2 and x_2's indirect effect on y_1 through y_2, $\beta_{12}\gamma_{22}\text{COV}(x_1, x_2)$, again multiplied by the $(1 - \beta_{12}\beta_{21})$ multiplier.

Calculating Direct, Indirect, and Total Effects

Researchers are largely interested in determining the direct, indirect, and total effects of one variable on another. For simple recursive models, the easiest method to determine direct, indirect, and total effects is through tracing paths in a path diagram. Looking at Figure 6.1, suppose we are interested in the effect of x_1 on y_2. The direct effect of x_1 is captured by the γ_{21} coefficient.

A researcher would rarely be interested in only interpreting the γ_{21} coefficient when discussing the effect of x_1 on y_2, but he or she would probably also like to discuss the indirect effect through y_1. This indirect effect can be computed by multiplying γ_{11} by β_{21}. This product shows the indirect effect of x_1 on y_2 as mediated by y_1. For this particular model, this is the entire indirect effect of x_1 on y_2. However, more complicated models may have additional indirect paths through which x_1 affects y_2.

Remember that there may also be interest in reporting the total effect in addition to interpreting the direct and indirect effects of x_1 on y_2. The total effect is simply the sum of the direct and indirect effects and is of substantive interest since it describes the extent to which y_2 would change, on average, if x_1 increases one unit. In some instances, researchers or policymakers may be less interested in specific indirect or direct effects that lead to a change in y_2 but more interested in the total change in y_2.

Tracing paths is an imperfect strategy for obtaining direct, indirect, and total effects in large and complicated recursive models and in nonrecursive models. Applying the insights from above, in the presence of reciprocal effects or feedback loops, direct, indirect, and total effects must be appropriately modified by the multiplier. For this reason, there are alternative and more precise ways to calculate direct and indirect effects.

The most accurate way to obtain direct, indirect, and total effects is based on manipulations of the matrices of the parameter estimates (Fox, 1980; Sobel, 1988). Consider the general case:

$$\mathbf{y} = \mathbf{B}\mathbf{y} + \mathbf{\Gamma}\mathbf{x} + \boldsymbol{\zeta}$$

Direct effects are drawn directly from this general matrix representation of simultaneous equation models.

$$\mathbf{D}_{yy} = \mathbf{B} \tag{6.6}$$

provides the direct effects of the endogenous variables on the endogenous variables, while

$$\mathbf{D}_{yx} = \mathbf{\Gamma} \tag{6.7}$$

provides the direct effects of the exogenous variables on the endogenous variables.

Indirect effects, denoted by matrix **N** in this presentation, are the sum of all paths of length two or longer. For a *recursive* model, the indirect effects between the endogenous variables are therefore denoted as follows:

$$\mathbf{N}_{yy} = \sum_{i=2}^{p-1} \mathbf{B}^i = \mathbf{B}^2 + \mathbf{B}^3 + \cdots + \mathbf{B}^{p-1} \tag{6.8}$$

where p is the number of endogenous variables and $\mathbf{B}^2 = \mathbf{BB}$. The expression does not include **B**, which is the matrix of direct effects, but instead includes all higher powers of the **B** matrix.

Indirect effects of the exogenous variables on the endogenous variables are denoted as follows:

$$\mathbf{N}_{yx} = \left(\sum_{i=1}^{p-1} \mathbf{B}^i\right)\mathbf{\Gamma} = \left(\mathbf{B}^1 + \mathbf{B}^2 + \cdots + \mathbf{B}^{p-1}\right)\Gamma \tag{6.9}$$

It is straightforward to compute indirect effects using these matrix powers for a recursive model. Total effects for recursive models can be similarly computed by adding direct effects to Equations 6.8 and 6.9:

$$\mathbf{T}_{yy} = \sum_{i=1}^{p-1} \mathbf{B}^i = \mathbf{B} + \mathbf{B}^2 + \mathbf{B}^3 + \cdots + \mathbf{B}^{p-1} \tag{6.10}$$

$$\mathbf{T}_{yx} = \left(\sum_{i=0}^{p-1} \mathbf{B}^i\right)\mathbf{\Gamma} = \mathbf{\Gamma} + \mathbf{B\Gamma} + \mathbf{B}^2\mathbf{\Gamma} + \mathbf{B}^3\mathbf{\Gamma} + \cdots + \mathbf{B}^{p-1}\mathbf{\Gamma} \tag{6.11}$$

Without feedback loops, in a recursive model the chain of influence captured by an indirect or total effect can be completely traced with Equations 6.8, 6.9, 6.10, and 6.11. But *nonrecursive* models are more complicated in that the model can imply an infinite chain of influence due to the presence of a reciprocal relationship or feedback loop.

$$\mathbf{T}_{yy} = \sum_{i=1}^{\infty} \mathbf{B}^i \tag{6.12}$$

As a consequence, $\mathbf{B}^i$ must converge to zero as $i \to \infty$ for the total effects to be defined. Otherwise, the system is not in equilibrium and indirect or total effects should not be described. Convergence requires that the absolute value of the largest eigenvalue of **B** be less than one (Bentler &

Freeman, 1983).[1] In the simple and common case of a model with a single loop, convergence occurs if $|\beta_{12}\beta_{21}| < 1$. If $|\beta_{12}\beta_{21}| > 1$, then the total and indirect effects are not defined (Bollen, 1987).

In short, the more appropriate way to calculate indirect and total effects for nonrecursive models is as follows:

$$\mathbf{D}_{yy} = \mathbf{B} \tag{6.13}$$

$$\mathbf{D}_{yx} = \mathbf{\Gamma} \tag{6.14}$$

$$\mathbf{N}_{yy} = (\mathbf{I} - \mathbf{B})^{-1} - \mathbf{I} - \mathbf{B} \tag{6.15}$$

$$\mathbf{N}_{yx} = (\mathbf{I} - \mathbf{B})^{-1}\mathbf{\Gamma} - \mathbf{\Gamma} \tag{6.16}$$

$$\mathbf{T}_{yy} = (\mathbf{I} - \mathbf{B})^{-1} - \mathbf{I} \tag{6.17}$$

$$\mathbf{T}_{yx} = (\mathbf{I} - \mathbf{B})^{-1}\mathbf{\Gamma} \tag{6.18}$$

Convergence and Equilibrium

As noted above, researchers should only investigate and interpret systems of equations that are in equilibrium—those that converge to Equations 6.13, 6.14, 6.15, 6.16, 6.17, and 6.18. To intuitively understand equilibrium, consider two different scenarios (Heise, 1975). Systems in an amplification relationship exhibit a positive/positive relationship between two variables. If the amplification is in equilibrium, then the effect of the positive feedback loop is ever increasing with a decreasing effect so that a single change to one of the two variables "settles down" to a single measureable indirect or total effect. An example of an amplification relationship in equilibrium would be the relationship between sales and advertising. An unstable amplification system suggests ever increasing cycles, with each cycle increasing until breakdown.

A second scenario would be a control scenario, representing a positive/negative relationship between two variables. In a control system, as Variable 1 increases, Variable 2 decreases. The decrease in Variable 2 causes Variable 1 to decrease, so that Variable 2 then increases. If the control relationship is in equilibrium, then the two variables should stay within a

---

[1] Goldberg (1958) and Bollen (1987) provide a sufficient but not necessary shortcut: If the elements of **B** are positive and the elements in the columns sum to less than one, the absolute values of the eigenvalues are less than one.

limited range of values. An example of a controlled system in equilibrium could be crime and law enforcement (Cornwell & Trumbull, 1994). Again, instability is possible. An unstable control system suggests oscillations of greater and greater magnitude.

Example Calculation of Direct, Indirect, and Total Effects: Recursive Example

Consider the following recursive model, which comes from the classic stratification model of Blau and Duncan (1967) and models the exogenous effects of a father's education and occupation on the respondent's education, first job, and current occupation (a complete description of computing these direct, indirect, and total effects can be found in Fox, 1980). We have

$$\mathbf{y} = \mathbf{B}\mathbf{y} + \mathbf{\Gamma}\mathbf{x} + \boldsymbol{\zeta}$$

where

$$\mathbf{\Gamma} = \begin{bmatrix} .31 & .279 \\ 0 & .224 \\ 0 & .115 \end{bmatrix}$$

and

$$\mathbf{B} = \begin{bmatrix} 0 & 0 & 0 \\ .44 & 0 & 0 \\ .394 & .281 & 0 \end{bmatrix}$$

Direct effects can be read directly from these two matrices. The coefficient matrix $\mathbf{\Gamma}$ (gamma) contains effects of exogenous variables on endogenous variables. So, for example, father's occupation (x_2) has a direct effect of .224 units on respondent's occupation (y_2), as shown by the γ_{22} coefficient in the gamma matrix. Coefficients describing the effect of an endogenous variable on another endogenous variable are summarized in the coefficient matrix, $\mathbf{B}$.

We could use either (6.8) or (6.15) to calculate the indirect effects of the endogenous variables on the endogenous variables. For example, using (6.8), we get

$$\mathbf{N}_{yy} = \sum_{i=2}^{p-1} \mathbf{B}^{i}$$

where $p - 1 = 2$ because there are three equations,

$$\mathbf{N}_{yy} = \begin{bmatrix} 0 & 0 & 0 \\ .44 & 0 & 0 \\ .394 & .281 & 0 \end{bmatrix} \begin{bmatrix} 0 & 0 & 0 \\ .44 & 0 & 0 \\ .394 & .281 & 0 \end{bmatrix} = \begin{bmatrix} 0 & 0 & 0 \\ 0 & 0 & 0 \\ .124 & 0 & 0 \end{bmatrix}$$

Indirect effects for exogenous on endogenous variables are computed using

$$\mathbf{N}_{yx} = \left(\sum_{i=1}^{2} \mathbf{B}^{i}\right)\mathbf{\Gamma} = (\mathbf{B}^{1} + \mathbf{B}^{2})\mathbf{\Gamma} = \mathbf{B}^{1}\mathbf{\Gamma} + \mathbf{B}^{2}\mathbf{\Gamma}$$

$$= \begin{bmatrix} 0 & 0 & 0 \\ .44 & 0 & 0 \\ .394 & .281 & 0 \end{bmatrix} \begin{bmatrix} .31 & .279 \\ 0 & .224 \\ 0 & .115 \end{bmatrix} + \begin{bmatrix} 0 & 0 & 0 \\ 0 & 0 & 0 \\ .124 & 0 & 0 \end{bmatrix} \begin{bmatrix} .31 & .279 \\ 0 & .224 \\ 0 & .115 \end{bmatrix}$$

$$= \begin{bmatrix} 0 & 0 \\ .136 & .123 \\ .161 & .207 \end{bmatrix}$$

Thus, for example, the indirect effect of father's occupation (x_2) on respondent's occupation (y_2) is .123.

The total effects, by extension, for exogenous variables on endogenous variables are

$$\mathbf{T}_{yx} = \mathbf{D}_{yx} + \mathbf{N}_{yx}$$

$$= \begin{bmatrix} .31 & .279 \\ 0 & .224 \\ 0 & .115 \end{bmatrix} + \begin{bmatrix} 0 & 0 \\ .136 & .123 \\ .161 & .207 \end{bmatrix} = \begin{bmatrix} .310 & .279 \\ .136 & .347 \\ .161 & .322 \end{bmatrix}$$

The total effect of father's occupation on respondent's occupation is .347 units, as shown in the total effects matrix, $\mathbf{T}_{yx}$. For endogenous variables on endogenous variables, the total effects are

$$\mathbf{T}_{yy} = \mathbf{D}_{yy} + \mathbf{N}_{yy}$$

$$= \begin{bmatrix} 0 & 0 & 0 \\ .44 & 0 & 0 \\ .394 & .281 & 0 \end{bmatrix} + \begin{bmatrix} 0 & 0 & 0 \\ 0 & 0 & 0 \\ .124 & 0 & 0 \end{bmatrix} = \begin{bmatrix} 0 & 0 & 0 \\ .44 & 0 & 0 \\ .518 & .281 & 0 \end{bmatrix}$$

Example Calculation of Direct, Indirect, and Total Effects: Nonrecursive Example

Consider the following nonrecursive model, which comes from the stratification study by Duncan, Haller, and Portes (1968) that studied the effect of father's education and occupation on the respondent's level of education and first job status. We have

$$\mathbf{y} = \mathbf{By} + \mathbf{\Gamma x} + \boldsymbol{\zeta}$$

where

$$\mathbf{\Gamma} = \begin{bmatrix} .27 & .15 & 0 & 0 \\ 0 & 0 & .16 & .35 \end{bmatrix}$$

and

$$\mathbf{B} = \begin{bmatrix} 0 & .40 \\ .34 & 0 \end{bmatrix}$$

Again, the direct effects can be pulled from the matrices **B** and **Γ**. Calculating indirect effects requires that we use Equations 6.15 and 6.16.

$$\begin{aligned} \mathbf{N}_{yy} &= (\mathbf{I} - \mathbf{B})^{-1} - \mathbf{I} - \mathbf{B} \\ &= \begin{bmatrix} 1 & .40 \\ .34 & 1 \end{bmatrix}^{-1} - \begin{bmatrix} 1 & 0 \\ 0 & 1 \end{bmatrix} - \begin{bmatrix} 0 & .40 \\ .34 & 0 \end{bmatrix} \\ &= \begin{bmatrix} 1.16 & .46 \\ .40 & 1.16 \end{bmatrix} - \begin{bmatrix} 1 & 0 \\ 0 & 1 \end{bmatrix} - \begin{bmatrix} 0 & .40 \\ .34 & 0 \end{bmatrix} \\ &= \begin{bmatrix} .16 & .065 \\ .055 & .16 \end{bmatrix} \end{aligned}$$

It is clear from $\mathbf{N}_{yy}$ that the endogenous variables can have indirect effects on themselves.

$$\begin{aligned} \mathbf{N}_{yx} &= (\mathbf{I} - \mathbf{B})^{-1}\mathbf{\Gamma} - \mathbf{\Gamma} \\ &= \begin{bmatrix} 1.16 & .46 \\ .40 & 1.16 \end{bmatrix} - \begin{bmatrix} .27 & .15 & 0 & 0 \\ 0 & 0 & .16 & .35 \end{bmatrix} \\ &\quad - \begin{bmatrix} .27 & .15 & 0 & 0 \\ 0 & 0 & .16 & .35 \end{bmatrix} \\ &= \begin{bmatrix} .04 & .02 & .073 & .17 \\ .11 & .06 & .03 & .06 \end{bmatrix} \end{aligned}$$

Total effects for this example can be calculated either by using Equations 6.17 and 6.18 or by adding direct and indirect effects.

Interpreting Theoretically Driven Specific Indirect Effects

The methods described previously enable the computation of overall indirect and total effects of variables on specific endogenous variables. There are instances, however, where for theoretical reasons a researcher may be interested in obtaining *specific indirect* effects. There are at least three types of effects transmitted by a particular variable or group of variables that can be distinguished (Bollen, 1987, pp. 50–52): (1) *exclusive-specific effects*, (2) *incremental-specific effects*, and (3) *inclusive-specific effects*. Each type of effect clarifies how a mediating variable operates within an overall model. Exclusive-specific effects focus only on a single direct path from one variable, through a mediating variable, to another variable (Greene, 1977). Incremental-specific effects focus on influences mediated by a variable of interest and all those after it (Alwin & Hauser, 1975). Inclusive-specific effects capture all compound paths working through a mediating variable (Fox, 1980).

To aid in understanding the difference between these effects, consider Figure 6.7, a six-variable model with three exogenous variables (x_1, x_2, and x_3) and three endogenous variables (y_1, y_2, and y_3).

Suppose we are theoretically interested in the indirect effects of x_1 on y_3 through y_2. The exclusive-specific effect focuses only on the single path through the mediating variable, y_2, that connects x_1 and y_3, resulting in an

Figure 6.7 Recursive Model With Multiple Indirect Effects

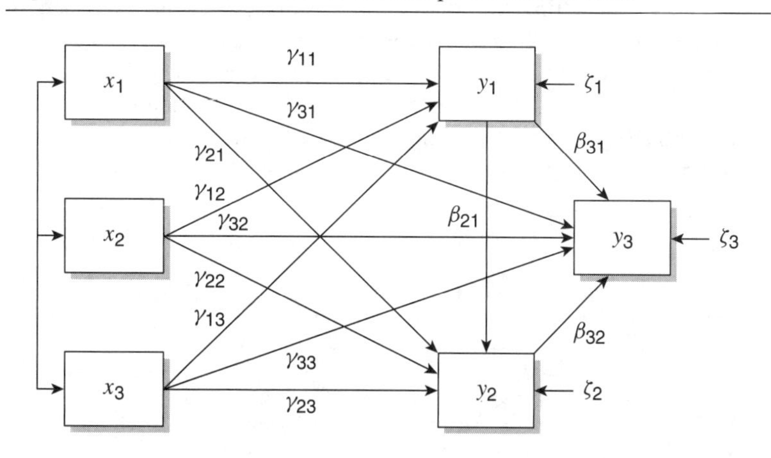

exclusive-specific effect of $\gamma_{21}\beta_{32}$. An exclusive-specific effect disregards other effects of x_1 on y_3 through variables in addition to y_2 (i.e., x_1 to y_1 to y_2 to y_3). To determine the incremental-specific effect, a researcher would first trace the path from x_1 to y_2 and then determine all possible paths that would reach y_3. In this example, there is only one path between y_2 and y_3, β_{32}, so the incremental-specific effect is identical to the exclusive-specific effect: $\gamma_{21}\beta_{32}$. Finally, the inclusive-specific effect includes all paths from x_1 to y_3 that pass through y_2. Looking at Figure 6.7 illustrates that in addition to the paths from x_1 to y_2 to y_3 implied by the other two approaches ($\gamma_{21}\beta_{32}$), the inclusive-specific effect will also include the path from x_1 to y_1 to y_2 to y_3 ($\gamma_{11}\beta_{21}\beta_{32}$), resulting in an inclusive-specific effect of ($\gamma_{11}\beta_{32}$) + ($\gamma_{11}\beta_{21}\beta_{32}$).

Using the same example model in Figure 6.7, consider the possible indirect effects of x_1 on y_3 mediated by y_1. In this case, the exclusive-specific effect is again a single path: the effect of x_1 on y_1 (γ_{11}) multiplied by the effect of y_1 on y_3 (β_{31}): $\gamma_{11}\beta_{31}$. For this mediated relationship, however, the incremental-specific effect is different. This approach first traces the effect from x_1 to y_1, and then determines all possible paths that can reach y_3. Thus, the incremental-specific effect not only includes the path from x_1 to y_1 to y_3 but also the path from x_1 to y_1 to y_2 to y_3: $\gamma_{11}\beta_{31} + \gamma_{11}\beta_{21}\beta_{32}$. For this example, the inclusive-specific effects are the same as the incremental-specific effects, since there are no other possible routes from x_1 to y_3 through y_1.

More on Mediation

Indirect effects are related to another topic of particular interest to many researchers—mediating variables. Researchers are often interested in understanding the mechanisms by which one variable influences another and model those mechanisms by theorizing and including mediating, or intervening, variables. For example, suppose a researcher is interested in understanding the relationship between education and fertility, as in Figure 6.8, where the coefficient from education to fertility is the total effect.

Figure 6.8 Total Effect of Education on Fertility

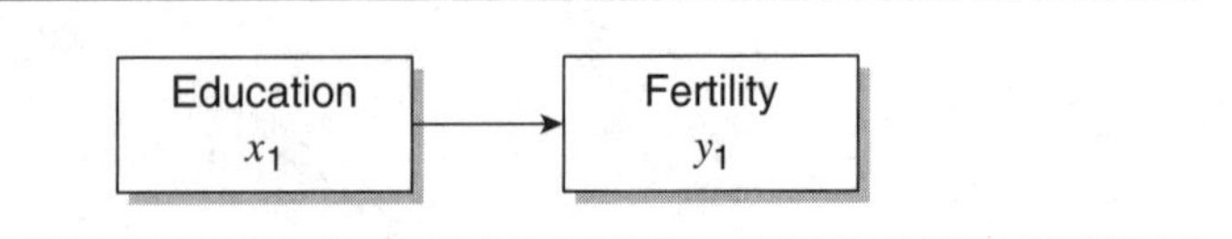

Theoretically, the researcher may want to understand how the relationship arises and, therefore, may envision three mediating variables that help explain the relationship between education and fertility: age at marriage, desired family size, and use of contraceptives, as shown in Figure 6.9.

Figure 6.9 Indirect Effects of Education on Fertility Through Three Mediating Variables

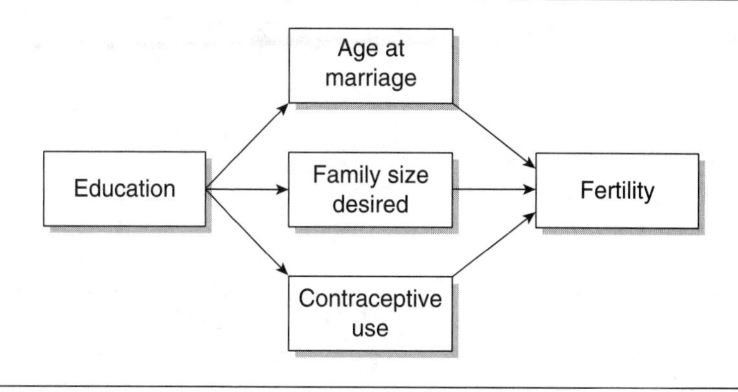

Figure 6.9 illustrates that the total effect is in fact composed of multiple indirect effects. That is, education has an indirect effect on fertility through the meditating variables.

Hypothesizing and testing mediating effects can be very useful, even in the case where there is no total effect between two variables. Consider suppressor effects—when direct and indirect effects are of opposite signs and similar magnitude, resulting in a nonsignificant total effect. In this situation, omitting the mediating variable would cause a suppressor effect. Bollen (1989b, p. 48) and MacKinnon, Krull, and Lockwood (2000, p. 175) recount an excellent example from McFatter (1979). Consider a study of assembly line workers that produces the counterintuitive finding that workers with higher intelligence do not make fewer mistakes than workers with lower intelligence, as in Figure 6.10.

Closer examination of the underlying process through a mediating variable, however, reveals that intelligence does reduce mistakes, but it also

Figure 6.10 Nonsignificant Total Effect

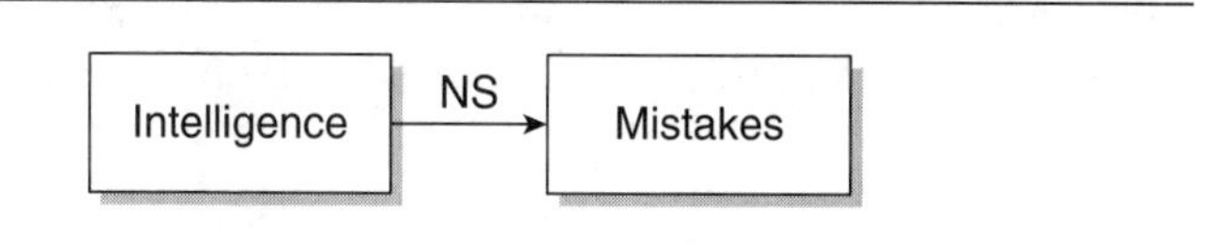

increases boredom that in turn increases mistakes, leading to a nonsignificant total effect (Figure 6.11).

Figure 6.11 Mediating Supressor Effect

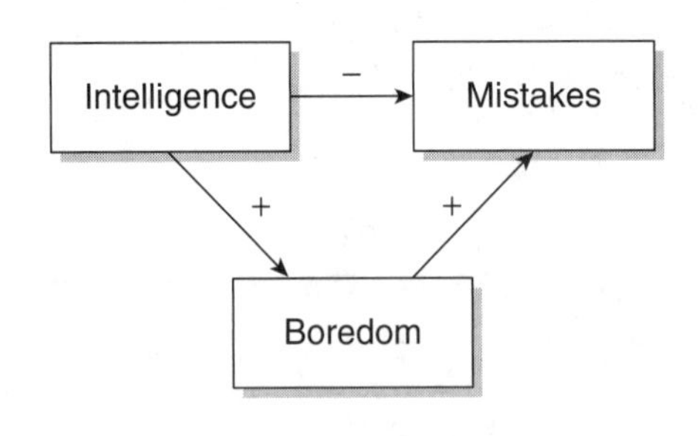

Researchers should be alert to the possibility of suppressor effects, which suggest that mediation can occur even if there is no significant relationship between the original exogenous and endogenous variables (Judd & Kenny, 1981; MacKinnon et al., 2000).

In modeling mediating effects, researchers may be interested in a single mediating variable, called "simple mediation" as in Figure 6.11, or they may be interested in multiple mediating variables, as in Figure 6.9 (Preacher & Hayes, 2008b). When multiple mediating variables are present, researchers are often interested in the total indirect effects as well as the exclusive-specific effects working through particular mediating variables.

Testing Indirect Effects: Simple Mediation

Beyond quantifying the effect size of an indirect effect, researchers may want to assess whether the indirect effects through mediating variables are statistically significant. In this section, we begin by addressing assessment and testing in the case of simple mediation—a single indirect effect working through a single mediating variable, as in Figure 6.12. Figure 6.12 presents the classic way to conceptualize mediation. Where c is the total effect of x on y, and c' is the direct effect of x on y partialing the indirect effect through mediator M.

Baron and Kenny (1986) provide an intuitive sense of how to establish that a mediated relationship exists. Consider the equation for the total effect of x on y,

$$y = cx + \zeta \tag{6.19}$$

Figure 6.12 Classic Conceptualization of a Simple Mediation Model

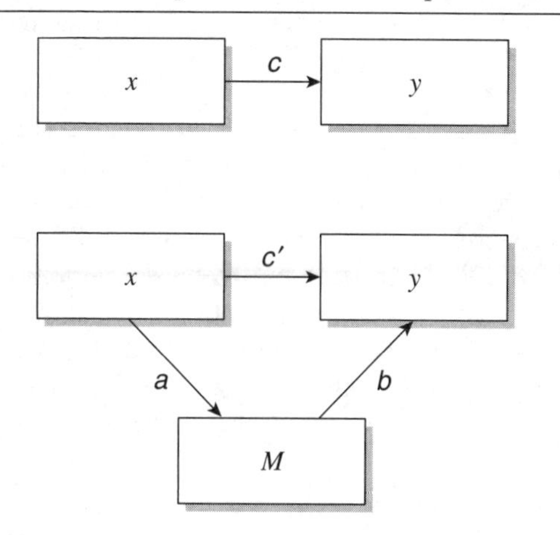

and the equations for the mediated effect of x on y:

$$M = ax + \zeta \tag{6.20}$$

$$y = c'x + bM + \zeta \tag{6.21}$$

Baron and Kenny (1986) require three criteria to declare that a relationship is mediated:

1. The independent variable must significantly affect the mediator in Equation 6.20.
2. The mediator must significantly affect the dependent variable in Equation 6.21.
3. The independent variable's previously significant effect on the dependent variable (c in Equation 6.19) is no longer significant once M is controlled (c' in Equation 6.21).

Criterion 3 describes a situation of *complete mediation*. Also possible is a situation of *partial mediation*, where the direct effect of x on y (c') is reduced after inclusion of M but remains statistically significant.

Although popular and intuitive, the Baron and Kenny method, sometimes called the causal steps strategy, does not directly test the significance of the indirect effect. As succinctly explained by Preacher and Hayes (2008b),

> Importantly, the causal steps strategy obliges the researcher to infer the presence and extent of mediation from a pattern of hypothesis tests,

> none of which directly addresses the hypothesis of interest—whether the causal path linking x to y through M is nonzero and in the direction expected. (p. 20)

A more direct test of an indirect effect builds and tests the product of the coefficients a and b: ab. A number of methods use ab as the basis for statistical inference about indirect effects. Remember that the point estimate for the indirect effect is the product of the coefficients: in this case, ab. For hypothesis testing, therefore, we must determine whether ab is significantly different from zero. Alternatively, we may want to report a confidence interval for the value of ab.

The most common test for simple mediation is a special case of the delta method (discussed below) called the Sobel test (Sobel, 1982):[2]

$$se_{ab} = \sqrt{b^2 s_a^2 + a^2 s_b^2 + s_a^2 s_b^2} \tag{6.22}$$

where s_a^2 is the asymptotic variance of a, and s_b^2 is the asymptotic variance of b. Some presentations omit the third term, given that it will often be a value close to zero. To test the indirect effect for significance, therefore, all that is required is to divide ab by se_{ab} and to compare the ratio with a standard normal distribution.

Testing Indirect Effects: Multivariate Delta Method

Researchers may be interested in testing multiple indirect effects or effects in more complicated models. To do so, the method is basically the same as that described above: We need to construct the sampling distribution of the indirect effects through determination of the asymptotic variance/covariance matrix.

If we view an indirect effect as a function of the direct effects between variables, then we can use the multivariate delta method, which allows us to obtain the asymptotic distribution for a differentiable function of a multinormal random vector.

In brief, the delta method generally states that if an estimator at our sample size N, $\hat{\theta}_N$, is consistent and normally distributed around θ with asymptotic variance/covariance matrix ACOV(θ), then in large samples, a function of $\hat{\theta}_N$, or $f(\hat{\theta}_N)$, is normally distributed with mean

$$f(\theta)$$

[2]The Sobel test is a variant of the multivariate delta method for the case of simple mediation. It assumes normality of the sampling distribution of ab. A simple Sobel test can be conducted using the command **sgmediation** after **regress** in Stata.

and asymptotic variance/covariance

$$\left(\frac{df}{d\theta}\right)' \text{ACOV}(\theta)\left(\frac{df}{d\theta}\right)$$

where $(df/d\hat{\theta})$ is the derivative of the function with respect to θ.

In short, using the delta method involves manipulating the asymptotic variance/covariance matrix of the original vector to get the asymptotic variance/covariance matrix of the function vector. In the case of indirect effects in simultaneous equations models, we have the asymptotic variance/covariance matrix for the direct effects and need to determine the asymptotic variance/covariance matrix for the indirect effects.

Consider a general case with d direct effects (θ) and i indirect effects of interest ($f(\theta)$). The large sample estimate of the asymptotic variance/covariance matrix of the indirect effects is

$$\left(\frac{df}{d\hat{\theta}_N}\right)' \text{ACOV}(\hat{\theta}_N)\left(\frac{df}{d\hat{\theta}_N}\right) \tag{6.23}$$

where $(df/d\hat{\theta}_N)$ is the first derivative with respect to θ with order $d \times i$, and $\text{ACOV}(\hat{\theta}_N)$ is the asymptotic covariance matrix of the direct effects with order $d \times d$. The resulting asymptotic covariance matrix of the indirect effects will have order $i \times i$.

To obtain $(df/d\theta)$, we stack the d direct effects into the vector θ and the i indirect effects into the vector $f(\theta)$. Each element of $f(\theta)$ is then differentiated with respect to the direct effects, resulting in a matrix of order $d \times i$:

$$\left(\frac{df}{d\theta}\right) = \begin{bmatrix} \frac{df_1}{d\theta_1} & \cdots & \frac{df_i}{d\theta_1} \\ \vdots & \ddots & \vdots \\ \frac{df_1}{d\theta_d} & \cdots & \frac{df_i}{d\theta_d} \end{bmatrix} \tag{6.24}$$

To illustrate use of the delta method, consider the model with three exogenous and three endogenous variables portrayed in Figure 6.7. Sobel (1982) provides the asymptotic standard errors for this example, beginning with the following direct effects (with standard errors in parentheses):

$$\begin{array}{ll} \gamma_{11} = .0385 & (.0025) \\ \gamma_{12} = .1707 & (.0156) \\ \gamma_{13} = -.2281 & (.0176) \end{array}$$

$$\gamma_{21} = .1352 \qquad (.0175)$$
$$\gamma_{22} = .0490 \qquad (.1082)$$
$$\gamma_{23} = -.4631 \qquad (.1231)$$

$$\beta_{21} = 4.3767 \qquad (.1202)$$

$$\gamma_{31} = .0114 \qquad (.0045)$$
$$\gamma_{32} = .0712 \qquad (.0275)$$
$$\gamma_{33} = -.0373 \qquad (.0314)$$

$$\beta_{31} = .1998 \qquad (.0364)$$
$$\beta_{32} = .0704 \qquad (.0045)$$

We begin with a simple example: testing the indirect effect $\beta_{21}\gamma_{11}$.

We have estimates of the direct effects and their standard errors: $\hat{\gamma}_{11} = .038$ with standard error .002 and $\hat{\beta}_{21} = 4.377$ with standard error .120. Furthermore, we can calculate an estimate of the indirect effect as $f(\hat{\theta}_N) = \hat{\beta}_{21}\hat{\gamma}_{11} = .168$. Stacking these into the two vectors results in

$$\theta = [\beta_{21} \ \gamma_{11}]' \tag{6.25}$$

and

$$f(\theta) = \beta_{21}\gamma_{11} \tag{6.26}$$

To calculate the standard error of the indirect effects, we need $(df/d\theta)$ and $\text{ACOV}(\hat{\theta}_N)$, an estimate of $\text{ACOV}(\theta)$:

$$\left(\frac{df}{d\theta}\right) = \left(\frac{df(\theta)}{d\beta_{21}} \ \ \frac{df(\theta)}{d\gamma_{11}}\right)' = [\gamma_{11} \ \beta_{21}]'$$

$$\text{ACOV}(\hat{\theta}_N) = \begin{bmatrix} s^2_{\hat{\beta}_{21}} & 0 \\ 0 & s^2_{\hat{\gamma}_{11}} \end{bmatrix}$$

Thus, the asymptotic variance/covariance

$$f(\hat{\theta}_N) = \left(\frac{df}{d\hat{\theta}_N}\right)' \text{ACOV}(\hat{\theta}_N)\left(\frac{df}{d\hat{\theta}_N}\right) \tag{6.27}$$

$$= [.038 \quad 4.377]\begin{bmatrix} .120^2 & 0 \\ 0 & .002^2 \end{bmatrix}\begin{bmatrix} .038 \\ 4.377 \end{bmatrix}$$

$$= .000097$$

making the standard error $= \sqrt{.000097} = .00987$

An alternative formula in the case of a simple two-path indirect effect is

$$\left[\hat{\beta}_{21}^2 \text{var}(\hat{\gamma}_{11}) + \hat{\gamma}_{11}^2 \text{var}(\hat{\beta}_{21})\right]^{1/2}$$

Note the similarity to the Sobel test.

Calculations using a larger number of indirect effects are also possible. Again using the example of the model in Figure 6.7, we can calculate six indirect effects on y_2 through y_1 and on y_3 through y_1: $\gamma_{11}\beta_{21}$, $\gamma_{12}\beta_{21}$, $\gamma_{13}\beta_{21}$, $\gamma_{11}\beta_{31}$, $\gamma_{12}\beta_{31}$, and $\gamma_{13}\beta_{31}$, which are functions of five direct effects: γ_{11}, γ_{12}, γ_{13}, β_{21}, and β_{31}. The resulting asymptotic covariance matrix will have order 6×6.

$$f(\hat{\theta}_N) = \left(\frac{df}{d\hat{\theta}_N}\right)' \text{ACOV}(\hat{\theta}_N)\left(\frac{df}{d\hat{\theta}_N}\right) = \begin{bmatrix} 4.38 & 0 & 0 & .039 & 0 \\ 0 & 4.38 & 0 & .17 & 0 \\ 0 & 0 & 4.38 & -.23 & 0 \\ .20 & 0 & 0 & 0 & .039 \\ 0 & .20 & 0 & 0 & .17 \\ 0 & 0 & .20 & 0 & -.23 \end{bmatrix}$$

$$\cdot \begin{bmatrix} .003^2 & 0 & 0 & 0 & 0 \\ 0 & .016^2 & 0 & 0 & 0 \\ 0 & 0 & .018^2 & 0 & 0 \\ 0 & 0 & 0 & .012^2 & 0 \\ 0 & 0 & 0 & 0 & .036^2 \end{bmatrix}$$

$$\cdot \begin{bmatrix} 4.38 & 0 & 0 & .20 & 0 & 0 \\ 0 & 4.38 & 0 & 0 & .20 & 0 \\ 0 & 0 & 4.38 & 0 & 0 & .20 \\ .039 & .17 & -.23 & 0 & 0 & 0 \\ 0 & 0 & 0 & .039 & .17 & -.23 \end{bmatrix}$$

Multiplication results in the asymptotic covariance matrix of the indirect effects. The square root of the diagonal produces the standard errors for the six indirect effects:

$$\begin{bmatrix} .012 & & & & & \\ & .071 & & & & \\ & & .082 & & & \\ & & & .001 & & \\ & & & & .007 & \\ & & & & & .009 \end{bmatrix}$$

It is also possible to use this method to calculate more complicated indirect effects, such as the effect of x_1 on y_3 through y_2, or $\gamma_{21}\beta_{32} + \gamma_{11}\beta_{21}\beta_{32}$ (Sobel, 1982). See Preacher and Hayes (2008a) for a discussion of the multivariate delta method in the special case of multiple mediation models.

Testing Indirect Effects: Bootstrapping

The Sobel test and the multivariate delta method assume normality of the coefficients, which may be of concern in small samples. Another strategy, bootstrapping, was first suggested in the context of indirect effects by Bollen and Stine (1992). It does not make distributional assumptions and may therefore be preferable in smaller sample sizes or when a researcher has reason to believe that nonnormality is present (Lockwood & MacKinnon, 1998; MacKinnon, Lockwood, & Williams, 2004; Preacher & Hayes, 2008a).

Bootstrapping indirect effects is similar to bootstrapping in other contexts. A researcher would sample from the original sample *with replacement* and obtain estimates of a and b. Sampling with replacement means that if a case is chosen for the bootstrapped sample, it remains in the pool to be potentially drawn again until sample size N is reached. A bootstrapped sample can therefore include the same case multiple times and not include some cases at all.

The process of resampling is repeated numerous (k) times, often more than 1,000 times, and for each new sample, a and b are estimated and used to calculate the indirect effect, ab. The distribution of these k values of ab then serves as an empirical approximation of the sampling distribution of ab. The mean of the empirical sampling distribution serves as the point estimate for ab, while the standard deviation of the empirical sampling distribution serves as the standard error. Confidence intervals can be obtained by sorting the resampled values of *ab* from low to high and defining appropriate upper and lower bounds. Note that these are percentiles and therefore need not be symmetric around the mean. The researcher can directly use the bootstrapped confidence intervals, use bias-corrected confidence intervals, or use bias-corrected and accelerated confidence intervals (Efron, 1987). The null hypothesis of no indirect effect can therefore be tested by assessing whether the confidence interval contains zero.

Empirical Example: Testing Indirect Effects

In Stata, one can compute the standard errors for indirect effects by using the delta method or by bootstrapping. To demonstrate this, we modify our running example to be a simple mediated model where (1) fear of walking in the neighborhood mediates the relationship between experiencing a robbery or burglary and interpersonal trust and (2) frequency of newspaper reading mediates the same relationship (see Figure 6.13).

Figure 6.13 Mediation Model, Interpersonal Trust Empirical Example

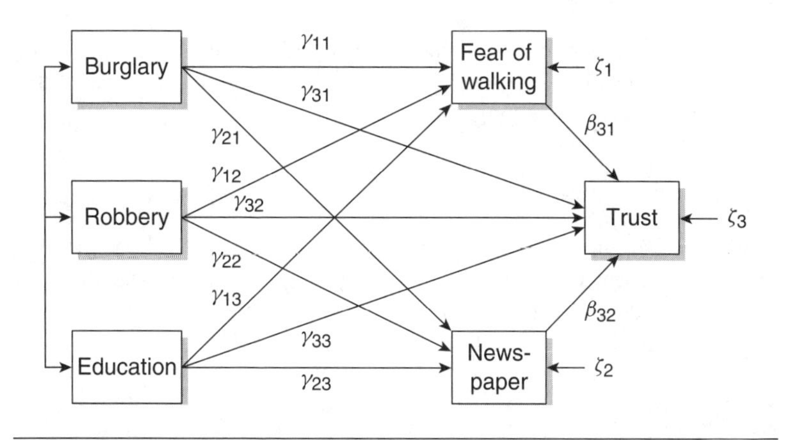

Rather than separately estimating each equation to obtain the coefficients, we can use the seemingly unrelated regression command (**sur**) to estimate all coefficients in the model simultaneously.

```
sureg (fearwalk burglary robbery educ)(news burglary robbery
educ)(intprtrst news fearwalk burglary robbery educ)
```

To compute the standard error of, for example, the indirect effect of burglary on interpersonal trust through fear of walking in the neighborhood, we use code for a nonlinear combination:

```
nlcom [fearwalk]_b[burglary]*[intprtrst]_b[fearwalk]
```

where the term in the first set of brackets refers to the outcome variable of the equation and "_ b" asks Stata to provide the estimated coefficient for the variable listed in the following set of brackets. Thus, here we are multiplying the burglary coefficient from the equation with fear of walking as the outcome ([fearwalk]_b[burglary]) by the fear of walking coefficient from the equation with interpersonal trust as the outcome ([intprtrst]_b[fearwalk]). The code to compute the standard error of the indirect effect of robbery on interpersonal trust is as follows:

```
nlcom [fearwalk]_b[robbery]*[intprtrst]_b[fearwalk]
```

The code to obtain and test the two indirect effects through newspaper readership is as follows:

```
nlcom [news]_b[burglary]*[intprtrst]_b[news]
nlcom [news]_b[robbery]*[intprtrst]_b[news]
```

And if we wished to test the inclusive-specific indirect effect operating through fear of walking by these two measures, we could use this code (that sums the two indirect paths):

```
nlcom [fearwalk]_b[burglary]*[intprtrst]_b[fearwalk] +
[fearwalk]_b[robbery]*[intprtrst]_b[fearwalk]
```

We can also estimate the standard errors by employing the bootstrapping approach advocated by Lockwood and MacKinnon (1998) and Preacher and Hayes (2008a) by defining a program called **bootmm**:

```
program bootmm, rclass
sureg (fearwalk burglary robbery educ)(news burglary robbery educ)(intprtrst
news fearwalk burglary robbery educ)
return scalar indbur = [fearwalk]_b[burglary]*[intprtrst]_b[fearwalk]
return scalar indrob = [fearwalk]_b[robbery]*[intprtrst]_b[fearwalk]
return scalar indtotal = [fearwalk]_b[burglary]*[intprtrst]_b[fearwalk]+ ///
[fearwalk]_b[robbery]*[intprtrst]_b[fearwalk]
return scalar indnewsbur = [news]_b[burglary]*[intprtrst]_b[news]
return scalar indnewsrob = [news]_b[robbery]*[intprtrst]_b[news]
end
```

This short program, called bootmm, is an **rclass** program because of the type of output we need to retrieve from these models (we are using the "return" command to obtain the coefficients). The **sureg** command estimates our model of interest, and the command lines after that tell Stata which coefficients we are interested in from this model. For example, the code

```
return scalar indbur = [fearwalk]_b[burglary]*[intprtrst]_b[fearwalk]
```

asks Stata to return a scalar value named "indbur" that is defined by the terms on the right-hand side of the equation (the indirect effect of burglary; defined as previously). The "///" allows lines to wrap, and "end" notes the end of the program.

We then ask Stata to compute these bootstrap results:

```
bootstrap r(indbur) r(indrob) r(indtotal) r(indnewsbur) r(indnewsrob),
reps(5000) nodots: bootmm2
```

This command computes the bootstrapped samples for the five coefficients of interest we defined above (indbur, indrob, indtotal, indnewsbur, and indnewsrob). It will perform 5,000 repetitions. We tell it to use the bootmm program we defined just above, and we ask it not to print a dot for each repetition ("nodots").

We can also ask for the percentiles from the bootstrapping (the option "*bc*" asks for the bias-corrected percentiles):

```
estat boot, bc percentile
```

For more information on programming indirect effects in Stata, see http://www.ats.ucla.edu/stat/stata/faq/mulmediation.htm.

Results of these models are presented in Table 6.1. The direct effects of the exogenous variables on fear of walking in the neighborhood are presented in the first column, the direct effects of the exogenous variables on reading a newspaper are presented in the second column, and the direct effects of the exogenous variables and the mediators on interpersonal trust are presented in the third column. Given that experiencing a burglary increases one's fear of walking in the neighborhood (.117) and fear of walking in the neighborhood decreases interpersonal trust (–.114), this indirect effect is –.013 ($.117 \times -.114 = -.013$), as shown in the first column of the lower panel. The first and second columns of the lower panel show that the estimated standard error for this indirect effect is similar whether we use the delta method or bootstrapping. In each case, we conclude that the indirect effect is significantly different from zero. Experiencing a burglary reduces the likelihood of reading a newspaper (–.234), and newspaper reading increases one's interpersonal trust (.070), again resulting in a significantly negative indirect effect on interpersonal trust (–.0162). And the indirect effect of robbery on interpersonal trust as mediated by fear of walking (–.0258), shown in the lower panel of Table 6.1, is significantly different from zero. However, the indirect effect of robbery on interpersonal trust working through newspaper reading is not significantly different from zero. We also see in the last two columns of the bottom panel that the bias-corrected bootstrapping confidence intervals for these three indirect effects do not include zero, further evidence that these effects are significantly different from zero. The only exception is for the indirect effect of experiencing a robbery, as mediated by reading a newspaper.

To estimate this model in SAS, we can use a macro based on Preacher and Hayes (2008a) and available for download from the web (http://www.comm.ohio-state.edu/ahayes/SPSS%20programs/indirect.htm). The code using the indirect macro for our model is as follows:

```
%indirect (data=a1,y=intprtrst,x=burglary,m=fearwalk news robbery educ
,c=2,boot=5000,conf=95, percent=1,bc=1,bca=1, normal=1);

%indirect (data=a1,y=intprtrst,x=robbery,m=fearwalk news burglary educ
,c=2,boot=5000,conf=95, percent=1,bc=1,bca=1, normal=1);
```

Each line of the code computes the mediation effect for a particular exogenous variable. For example, the first line of code tells SAS to use the indirect.mac macro, to use the a1 dataset, that the ultimate outcome variable (y) is intprtrst, and that the x variable for which to compute the indirect effects is burglary; the m= command tells SAS which variables are the mediators and which are additional control variables in the model (the c=2 command tells SAS that the last two variable names on this list are control variables—robbery and educ—and therefore the first two names on the list are mediators—fearwalk and news), percent=1 asks SAS to compute the bootstrapping confidence intervals (CIs), bc=1 asks SAS to compute the bias-corrected CIs, bca=1 asks for the bias-corrected and accelerated CIs, and normal=1 asks SAS to compute the Sobel standard errors.

Because this code only allows for a single x variable when computing these mediating effects, the second line of code instructs SAS to compute the standard errors and confidence intervals for the mediated indirect

Table 6.1 Results for Mediation Model: Computing Standard Errors for Indirect Effects With Two Different Approaches, Delta Method and Bootstrapping

	Direct Effects		
	Outcome: Fear of Walking in Neighborhood	*Outcome: Frequency Reading the Newspaper*	*Outcome: Interpersonal Trust*
Fear of walking in the neighborhood			–0.114 (0.023)
Frequency reading the newspaper			0.070 (0.010)
Experienced a burglary in the past year	0.117 (0.029)	–0.234 (0.068)	–0.269 (0.045)
Experienced a robbery in the past year	0.227 (0.048)	–0.022 (0.115)	–0.127 (0.076)
Years of education	–0.009 (0.002)	0.078 (0.006)	0.054 (0.004)
Intercept	0.562 (0.033)	2.007 (0.078)	–0.908 (0.057)
R^2	0.013	0.041	0.078

	Indirect Effects			
	Using Delta Method for Standard Errors	*Using Bootstrapping for Standard Errors*	*Bias-Corrected Bootstrapping 95% Confidence Intervals*	
Fear of walking mediating experienced a burglary	–0.0134 (0.0043)	–0.0134 (0.0043)	–0.0240	–0.0066
Fear of walking mediating experienced a robbery	–0.0258 (0.0076)	–0.0258 (0.0074)	–0.0436	–0.0138
Reading the newspaper mediating experienced a burglary	–0.0162 (0.0053)	–0.0162 (0.0055)	–0.0283	–0.0067
Reading the newspaper mediating experienced a robbery	–0.0015 (0.0080)	–0.0015 (0.0083)	–0.0188	0.0143

effect of robbery. Note that in this code, robbery is listed as the *x* variable (instead of burglary), and burglary is listed as one of the "control" variables (instead of robbery). Everything else remains the same.

The bootstrapped results from the SAS macro are very similar to those obtained from Stata.

CHAPTER 7. CONCLUSION

This monograph has focused on the important issues surrounding nonrecursive simultaneous equation models. We described how simultaneous equation models can be divided into two major types—recursive and nonrecursive—and focused on nonrecursive models, given the additional challenges they pose to a researcher. We blended two complementary perspectives on nonrecursive models: that of the structural equation modeling (SEM) with latent variables literature (e.g., Bollen, 1989b; Kaplan, 2009) and the econometrics tradition (e.g., Greene, 2008; Kennedy, 2008; Wooldridge, 2009).

This monograph described five key steps of simultaneous equations models: specification, identification, estimation, assessment, and interpretation. We emphasized that researchers should use prior theory in model specification to detail a series of equations representing their model. We stressed that identification is a crucial process, whereby researchers theoretically establish that a unique solution exists for each parameter of the model. The constraints imposed on model parameters to achieve identification should be guided by theory and should then be assessed, to the extent possible, with the techniques described in this monograph. Although many SEM texts focus almost exclusively on a full-information estimator such as maximum likelihood (ML), we stressed that a limited-information estimator such as two-stage least squares (2SLS) can be advantageous in many circumstances. In terms of assessment, we suggest that it requires (1) assessing the component fit of each individual equation, (2) assessing overall fit of the system of equations in the tradition of SEM with latent variables, and (3) assessing the quality of the instrumental variables. Finally, interpretation of simultaneous equations models is more complicated than single-equation techniques as researchers are able to interpret the direct, indirect, and total effects of one variable on another. We described techniques to compute these three types of effects, as well as methods to assess their statistical significance.

One key theme of this monograph is its emphasis that researchers rely on theory when specifying nonrecursive models. Theory is necessary for determining which paths should or should not be estimated. Furthermore, theory is crucial for the selection of instrumental variables, which are necessary for the identification and successful estimation of such models. In some instances when empirical tests of certain assumptions are not possible, existing theory will be all that the researcher has to justify choices in model specification.

A second key theme of this monograph is the importance of testing as many of the assumptions of nonrecursive models as possible. As we have demonstrated here, these tests are not particularly difficult to implement and can be conducted in most standard statistical software packages (e.g., SAS, Stata). Furthermore, the tests provide evidence that is crucially important for assessing the quality of the estimates. An unfortunate state of affairs is that researchers infrequently employ these tests. It may be that researchers overlook these tests because many SEM software programs using an ML estimator do not provide them. As a consequence, some researchers may assume that such tests are not necessary when using such an ML estimator. This is not true, as bad or weak instrumental variables undermine the properties of any estimator.

Given that estimation of nonrecursive systems hinges on instrumental variables, it is critical to assess the quality of the instrumental variables used in a model. We have highlighted that two key assumptions are made about IVs: (1) The instrumental variables are not correlated with the disturbance term and (2) the instrumental variables are correlated with the problematic variable (they are reasonably strong predictors of the problematic variable). Assumption (1) is about validity and can be assessed with the tests of overidentifying restrictions that we described. Assumption (2) can lead to inefficiency and increased finite sample bias and can be assessed with the tests for weak instruments that we described.

We have also taken the position throughout this monograph that limited-information estimators such as 2SLS are not outdated methods that can be safely ignored by researchers using structural equation software packages. In some instances, the assumptions of the 2SLS estimator may be more palatable to a researcher than those of the ML estimator. Full-information estimators may produce more efficient parameter estimates in a limited set of circumstances, but they are more sensitive to specification error, since error in one equation can be spread to parameter estimation in all equations of the system. Limited-information estimators such as 2SLS do not spread specification error among equations as long as they do not affect the excluded variables/instruments. Furthermore, the 2SLS estimator—with its explicit estimation of the first-stage equation in nonrecursive models—provides useful information of which researchers using an ML estimator may not be aware. We suggest that regardless of which estimation techniques researchers use, it is still important to obtain and present information about the first-stage equations to readers and to better assess the quality of the estimation.

REFERENCES

Akaike, H. (1974). A new look at the statistical model identification. *IEEE Transactions on Automation Control, AC-19,* 716–723.

Alwin, D. F., & Hauser, R. M. (1975). The decomposition of effects in path analysis. *American Sociological Review, 40,* 37–47.

Anderson, J., & Gerbing, D. W. (1984). The effects of sampling error on convergence, improper solutions and goodness-of-fit indices for maximum likelihood confirmatory factor analysis. *Psychometrika, 49,* 155–173.

Anderson, T. W., & Rubin, H. (1949). Estimation of the parameters of a single equation in a complete system of stochastic equations. *Annals of Mathematical Statistics, 20,* 46–63.

Andrews, D. W. K., Moreira, M. J., & Stock, J. H. (2006). Optimal two-sided invariant similar tests for instrumental variables regression. *Econometrica, 74*(3), 715–752.

Andrews, D. W. K., & Stock, J. H. (2007). Testing with many weak instruments. *Journal of Econometrics, 138*(1), 24–46.

Ansolabehere, S., & Jones, P. E. (2010). Constituents' responses to congressional roll-call voting. *American Journal of Political Science, 54*(3), 583–597.

Asher, H. B. (1983). *Causal modeling*. Beverly Hills, CA: Sage.

Baron, R., & Kenny, D. (1986). The moderator-mediator variable distinction in social psychological research: Conceptual, strategic, and statistical considerations. *Journal of Personality and Social Psychology, 51*(6), 1173–1182.

Barro, R. J., & McCleary, R. M. (2003). Religion and economic growth across countries. *American Sociological Review, 68,* 760–781.

Basmann, R. L. (1960). On finite sample distributions of generalized classical linear identifiability test statistics. *Journal of the American Statistical Association, 55,* 650–659.

Baum, C., Schaffer, M., & Stillman, S. (2003). Instrumental variables and GMM: Estimation and testing. *Stata Journal, 3*(1), 1–31.

Belsley, D., Kuh, E., & Welsch, R. (2004). *Regression diagnostics: Identifying influential data and sources of collinearity*. New York: Wiley.

Bentler, P. M. (1990). Comparative fit indexes in structural models. *Psychological Bulleting, 107*(2), 238–246.

Bentler, P., & Bonnett, D. (1980). Significance tests and goodness of fit in the analysis of covariance structures. *Psychological Bulletin, 88*(3), 588–606.

Bentler, P. M., & Freeman, E. H. (1983). Tests for stability in linear structural equation systems. *Psychometrika, 48*(1), 143–145.

Bentler, P. M., & Raykov, T. (2000). On measures of explained variance in nonrecursive structural equation models. *Journal of Applied Psychology, 85,* 125–131.

Berry, W. (1984). *Nonrecursive causal models*. Beverly Hills, CA: Sage.

Blau, P. M., & Duncan, O. D. (1967). *The American occupational structure.* New York: Wiley.

Bollen, K. A. (1987). Total, direct, and indirect effects in structural equation models. In C. C. Clogg (Ed.), *Sociological methodology 1987* (pp. 37–69). Washington, DC: American Sociological Association.

Bollen, K. A. (1989a). A new incremental fit index for general structural equation models. *Sociological Methods and Research, 17*(3), 303–316.

Bollen, K. A. (1989b). *Structural equations with latent variables*. New York: Wiley.

Bollen, K. A. (1996). An alternative two stage least squares (2SLS) estimator for latent variable equations. *Psychometrika, 61*(1), 109–121.

Bollen, K. A., Kirby, J. B., Curran, P. J., Paxton, P., & Chen, F. (2007). Latent variable models under misspecification two-stage least squares (2SLS) and maximum likelihood (ML) estimators. *Sociological Methods & Research, 36,* 48–86.

Bollen, K. A., & Kmenta, J. (1986). Estimation of simultaneous equation models with autoregressive or heteroscedastic disturbances. In J. Kmenta (Ed.), *Elements of econometrics* (pp. 704–711). New York: Macmillan.

Bollen, K. A., & Stine, R. A. (1992). Bootstrapping goodness-of-fit measures in structual equation models. *Sociological Methods and Research, 21*(2), 205–229.

Bollen, K. A., & Ting, K.-F. (1993). Confirmatory tetrad analysis. In P. V. Marsden (Ed.), *Sociological methodology* (Vol. 23, pp. 147–176). Cambridge, MA: Blackwell.

Bollen, K. A., & Ting, K.-F. (1998). Bootstrapping a test statistic for vanishing tetrads. *Sociological Methods & Research, 27,* 77–102.

Bollen, K. A., & Ting, K.-F. (2000). A tetrad test for causal indicators. *Psychological Methods, 5,* 3–22.

Bound, J., Jaeger, D. A., & Baker, R. M. (1995). Problems with instrumental variables estimation when the correlation between the instruments and the endogenous explanatory variable is weak. *Journal of the American Statistical Association, 90,* 443–450.

Brehm, J., & Rahn, W. (1997). Individual-level evidence for the causes and consequences of social capital. *American Journal of Political Science, 41,* 999–1023.

Browne, M. W. (1984). Asymptotically distribution-free methods for the analysis of covariance structures. *British Journal of Mathematical and Statistical Psychology, 37,* 62–83.

Browne, M. W., & Cudeck, R. (1993). Alternative ways of assessing model fit. In K. Bollen & J. S. Long (Eds.), *Testing structural equation models* (pp. 136–162). Newbury Park, CA: Sage.

Chen, F., Curran, P., Bollen, K., Kirby, J., & Paxton, P. (2008). An empirical assessment of the use of fixed cutoff points in RMSEA test statistic in structural equation models. *Sociological Methods and Research, 36*(4), 462–494.

Claibourn, M. P., & Martin, P. S. (2000). Trusting and joining? An empirical test of the reciprocal nature of social capital. *Political Behavior, 22,* 267–291.

Cornwell, C., & Trumbull, W. (1994). Estimating the economic model of crime with panel data. *Review of Economic Studies, 76,* 360–366.

Cragg, J. G. (1968). Some effects of incorrect specification on the small-sample properties of several simultaneous-equation estimators. *International Economic Review, 9,* 63–86.

Curran, P. J. (1994). *The robustness of confirmatory factor analysis to model misspecification and violations of normality.* Unpublished doctoral dissertation, Arizona State University, Tempe.

Curran, P. J., Bollen, K. A., Chen, F., Paxton, P., & Kirby, J. (2003). Finite sampling properties of the point estimates and confidence intervals of the RMSEA. *Sociological Methods and Research, 32*(2), 208–252.

Curran, P. J., West, S., & Finch, J. (1996). The robustness of test statistics to nonnormality and specification error in confirmatory factor analysis. *Psychological Methods, 1,* 16–29.

Cutler, D. M., & Glaeser, E. L. (1997). Are ghettos good or bad? *Quarterly Journal of Economics, 112,* 827–872.

Duncan, O. D. (1975). *Introduction to structural equation models.* New York: Academic Press.

Duncan, O. D., Featherman, D., & Duncan, B. (1972). *Socioeconomic background and achievement.* New York: Seminar Press.

Duncan, O. D., Haller, A. O., & Portes, A. (1968). Peer influences on aspirations: A reinterpretation. *American Journal of Sociology, 74*(2), 119–137.

Durbin, J. (1954). Errors in variables. *Review of the International Statistical Institute, 22,* 23–32.

Ebbes, P., Wedel, M., Böckenholt, U., & Steerneman, T. (2005). Solving and testing for regressor-error (in)dependence when no instrumental variables are available: With new evidence for the effect of education on income. *Quantitative Marketing and Economics, 3,* 365–392.

Efron, B. (1987). Better bootstrap confidence intervals. *Journal of the American Statistical Association, 82,* 171–185.

Ethington, C. A., & Wolfle, L. M. (1986). A structural model of mathematics achievement for men and women. *American Educational Research Journal, 23,* 65–75.

Fisher, F. M. (1961). On the cost of approximate specification in simultaneous equation estimation. *Econometrica, 29,* 139–170.

Fox, J. (1980). Effect analysis in structural equation models. *Sociological Methods and Research, 9*(1), 3–28.

Fox, J. (1991). *Regression diagnostics: An introduction.* Newbury Park, CA: Sage.

Fox, J. (2009). *A mathematical primer for social statistics* (T. F. Liao, Ed., Vol. 159). Thousand Oaks, CA: Sage.

Fukuyama, F. (1995). *Trust: The social virtues and the creation of prosperity.* New York: Free Press.

Fuller, W. (1977). Some properties of a modification of the limited information estimator. *Econometrica, 45,* 939–953.

Gill, J. (2006). *Essential mathematics for political and social research.* Cambridge, UK: Cambridge University Press.

Goldberg, S. (1958). *Introduction to difference equations.* New York: Wiley.

Greene, V. L. (1977). An algorithm for total and indirect causal effects. *Political Methodology, 4,* 369–381.

Greene, W. H. (2008). *Econometric analysis* (6th ed.). Upper Saddle River, NJ: Prentice Hall.

Guggenberger, P. (2010). The impact of a Hausman pretest on the asymptotic size of a hypothesis test. *Econometric Theory, 26,* 369–384.

Hahn, J., & Hausman, J. (2002). A new specification test for the validity of instrumental variables. *Econometrica, 70*(1), 163–189.

Hahn, J., Hausman, J., & Kuersteiner, G. (2004). Estimation with weak instruments: Accuracy of higher-order bias and MSE approximations. *Econometrics Journal, 7*(1), 272–306.

Hansen, L. (1982). Large sample properties of generalized method of moments estimators. *Econometrica, 50*(3), 1029–1054.

Hausman, J. A. (1978). Specification tests in econometrics. *Econometrica, 46,* 1251–1271.

Hausman, J. A. (1983). Specification and estimation of simultaneous equation models. In Z. Griliches & M. D. Intriligator (Eds.), *Handbook of econometrics* (Vol. 1, pp. 392–448). New York: North-Holland.

Hayduk, L. A. (2006). Blocked-error-R^2: A conceptually improved definition of the proportion of explained variance in models containing loops or correlated residuals. *Quality and Quantity, 40,* 629–649.

Heise, D. R. (1975). *Causal analysis.* New York: Wiley.

Hipp, J. R., & Bauer, D. J. (2002). CTANEST1: Program for testing nested and categorical tetrads. Retrieved December 3, 2010, from https://webfiles.uci.edu/hippj/johnhipp/ctanest1.htm

Hipp, J. R., Bauer, D. J., & Bollen, K. A. (2005). Conducting tetrad tests of model fit and contrasts of tetrad-nested models: A new SAS macro. *Structural Equation Modeling, 12,* 76–93.

Hipp, J. R., & Bollen, K. A. (2003). Model fit in structural equation models with censored, ordinal, and dichotomous variables: Testing vanishing tetrads. *Sociological Methodology, 33,* 267–305.

Hoxby, C. M. (1996). How teachers' unions affect education production. *Quarterly Journal of Economics, 111,* 671–718.

Hoxby, C. M. (2001). All school finance equalizations are not created equal. *Quarterly Journal of Economics, 116,* 1189–1231.

Hu, L.-T., & Bentler, P. M. (1995). Evaluating model fit. In R. Hoyle (Ed.), *Structural equation modeling: Concepts, issues, and applications.* Thousand Oaks, CA: Sage.

Hu, L.-T., & Bentler, P. M. (1999). Cutoff criteria for fit indexes in covariance structure analysis: Conventional criteria versus new alternatives. *Structural Equation Modeling, 6*(1), 1–55.

Jeong, J., & Yoon, B. H. (2010). The effect of pseudo-exogenous instrumental variables on Hausman test. *Communications in Statistics—Simulation and Computation, 39,* 315–321.

Jöreskog, K. G. (1999). *What is the interpretation of R^2?* Retrieved June 2001 from http://www.ssicentral.com/lisrel/techdocs/r2rev.pdf

Joreskog, K. and D. Sorbom (1986). LISREL VI: Analysis of Linear Structural Relationships by Maximum Likelihood and Least Square Methods. Mooresville, IN: Scientific Software.

Judd, C., & Kenny, D. (1981). Process analysis: Estimation mediation in treatment evaluations. *Evaluation Review, 5,* 602–619.

Kaplan, D. (1988). The impact of specification error on the estimation, testing, and improvement of structural equation models. *Multivariate Behavioral Research, 23,* 69–86.

Kaplan, D. (2009). *Structural equation modeling: Foundations and extensions.* Thousand Oaks, CA: Sage.

Kelejian, H. H. (1971). Two-stage least squares and econometric systems linear in parameters but nonlinear in the endogenous variables. *Journal of the American Statistical Association, 66,* 373–374.

Kennedy, P. (2008). *A guide to econometrics* (6th ed.). Malden, MA: Blackwell.

Kirby, J. B., & Bollen, K. A. (2009). Using instrumental variable tests to evaluate model specification in latent variable structural equation models. *Sociological Methodology, 39*(1), 327–355.

Kmenta, J. (1997). *Elements of econometrics* (2nd ed.). Ann Arbor: University of Michigan Press.

Kritzer, H. M. (1984). Mothers and fathers and girls and boys: Socialization in the family revisited. *Political Methodology, 10,* 245–265.

Levitt, S. D. (1996). The effect of prison population size on crime rates: Evidence from prison overcrowding litigation. *Quarterly Journal of Economics, 111,* 319–351.

Liska, A., & Bellair, P. (1995). Violent crime rates and racial composition: Convergence over time. *American Journal of Sociology, 101*(3), 578–610.

Lockwood, C., & MacKinnon, D. (1998). Bootstrapping the standard error of the mediated effect. In *Proceedings of the 23rd Annual SAS Users Group International Conference* (pp. 997–1002). Cary, NC: SAS Institute.

Long, J. S. (1988). *Common problems/proper solutions: Avoiding errors in quantitative research.* Newbury Park, CA: Sage.

MacKinnon, D., Krull, J., & Lockwood, C. (2000). Equivalence of the mediation, confounding and suppression effect. *Prevention Science, 1*(4), 173–181.

MacKinnon, D., Lockwood, C., & Williams, J. (2004). Confidence limits for the indirect effect: Distribution of the product and resampling methods. *Multivariate Behavior Research, 39*(1), 99–128.

Magdalinos, M. A., & Symeonides, S. D. (1996). A reinterpretation of the tests of overidentifying restrictions. *Journal of Econometrics, 73,* 325–353.

Markowitz, F. E., Bellair, P. E., Liska, A. E., & Liu, J. (2001). Extending social disorganization theory: Modeling the relationships between cohesion, disorder, and fear. *Criminology, 39,* 293–319.

McFatter, R. M. (1979). The use of structural equation models in interpreting regression equations including suppressor and enhancer variables. *Applied Psychological Measurement, 3,* 123–135.

Moreira, M. (2003). A conditional likelihood test for structural models. *Econometrica, 71*(4), 1027–1048.

Murray, M. P. (2006a). Avoiding invalid instruments and coping with weak instruments. *Journal of Economic Perspectives, 20*(4), 111–132.

Murray, M. P. (2006b). *Econometrics: A modern introduction.* Boston: Pearson.

Namboodiri, K. (1984). *Matrix algebra: An introduction* (R. G. Niemi, Ed., Vol. 38). Beverly Hills, CA: Sage.

Paxton, P. (1999). Is social capital declining in the United States? A multiple indicator assessment. *American Journal of Sociology, 105,* 88–127.

Paxton, P. (2002). Social capital and democracy: An interdependent relationship. *American Sociological Review, 67*(2), 254–277.

Paxton, P. (2007). Association memberships and generalized trust: A multilevel model across 31 countries. *Social Forces, 86,* 47–76.

Preacher, K., & Hayes, A. (2008a). Asymptotic and resampling strategies for assessing and comparing indirect effects in multiple mediator models. *Behavior Research Methods, 40*(3), 879–891.

Preacher, K., & Hayes, A. (2008b). Contemporary approaches to assessing mediation in communication research. In A. Hayes, M. Slater, & L. Snyder (Eds.), *The SAGE sourcebook of advanced data analysis methods for communication research* (pp. 13–54). Thousand Oaks, CA: Sage.

Putnam, R. D. (1993). *Making democracy work: Civic traditions in modern Italy*. Princeton, NJ: Princeton University Press.

Raftery, A. (1995). Bayesian model selection in social research. *Sociological Methodology, 25,* 111–163.

Rigdon, E. E. (1995). A necessary and sufficient identification rule for structural equation models estimated in practice. *Multivariate Behavioral Research, 30*(3), 359–383.

Sadler, P., & Woody, E. (2003). Is who you are who you're talking to? Interpersonal style and complementarity in mixed-sex interactions. *Journal of Personality and Social Psychology, 84,* 80–96.

Sargan, J. D. (1958). The estimation of economic relationships using instrumental variables. *Econometrica, 26,* 393–415.

Schwarz, G. (1978). Estimating the dimension of a model. *Annals of Statistics, 6*(2), 461–464.

Shah, D. V. (1998). Civic engagement, interpersonal trust, and television use: An individual-level assessment of social capital. *Political Psychology, 19,* 469–496.

Shea, J. (1997). Instrumental relevance in multivariate linear models: A simple measure. *Review of Economics and Statistics, 79,* 348–352.

Sobel, M. (1982). Asymptotic confidence intervals for indirect effects in structural equation models. *Sociological Methodology, 13*, 290–312.

Sobel, M. E. (1988). Direct and indirect effect in linear structural equation models. In J. S. Long (Ed.), *Common problems/proper solutions: Avoiding error in quantitative research* (pp. 46–64). Newbury Park, CA: Sage.

Staiger, D., & Stock, J. H. (1997). Instrumental variables regression with weak instruments. *Econometrica, 65,* 557–586.

Steiger, J. H., & Lind, J. C. (1980, May). *Statistically-based tests for the number of common factors*. Paper presented at the annual spring meeting of the Psychometric Society, Iowa City, IA.

Stock, J. H., & Yogo, M. (2005). Testing for weak instruments in linear IV regression. In J. H. Stock & D. W. K. Andrews (Eds.), *Identification and inference for econometric models:*

A festschrift in honor of Thomas Rothenberg (pp. 80–108). Cambridge, UK: Cambridge University Press.

Tanaka, J. S. (1993). Multifaceted conceptions of fit in structural equation models. In K. Bollen & J. S. Long (Eds.), *Testing structural equation models* (pp. 10–39). Newbury Park, CA: Sage.

Teel, J. E., Jr., Bearden, W. O., & Sharma, S. (1986). Interpreting LISREL estimates of explained variance in nonrecursive structural equation models. *Journal of Marketing Research, 23,* 164–168.

Tucker, L., & Lewis, C. (1973). A reliability coefficient for maximum likelihood factor analysis. *Psychometrika, 38*(1), 1–10.

Waite, L. J., & Stolzenberg, R. M. (1976). Intended childbearing and labor force participation of young women: Insights from nonrecursive models. *American Sociological Review, 41,* 235–251.

Wooldridge, J. (2002). *Econometric analysis of cross section and panel data*. Cambridge: MIT Press.

Wooldridge, J. M. (2009). *Introductory econometrics: A modern approach* (4th ed.). Cincinnati, OH: South-Western.

Wu, D. M. (1974). Alternative tests of independence between stochastic regressors and disturbances: Finite sample results. *Econometrica, 42*(3), 529–546.

Young, C. (2009). Model uncertainty in sociological research: An application to religion and economic growth. *American Sociological Review, 74,* 380–397.

Zellner, A. (1962). An efficient method of estimating seemingly unrelated regression equations and tests for aggregation. *Journal of the American Statistical Association, 57*(297), 348–368.

Zellner, A., & Theil, H. (1962). Three-stage least squares: Simultaneous estimation of simultaneous equations. *Econometrica, 30*(1), 54–78.

AUTHOR INDEX

SUBJECT INDEX